AN INTRODUCTION
TO THE CORIOLIS FORCE

GASPARD GUSTAVE de CORIOLIS
1792–1843

Coriolis entered the École Polytechnique in 1808. After graduation he served for several years in the corps of engineers (of the Ponts et Chauseés). In 1816 he started his teaching career. He was an attentive, effective and solicitous teacher. He was also a bachelor and invalid. As a result of studying formulations of dynamical problems in rotating machinery he was led to consider the effect of changes of coordinate systems in analytical mechanics. The result of these studies was presented to the Académie des Sciences on June 6, 1831.

A French oceanographic vessel was named after him in 1963.

AN INTRODUCTION TO THE CORIOLIS FORCE

Henry M. Stommel

Dennis W. Moore

COLUMBIA UNIVERSITY PRESS
New York

Columbia University Press
New York   Oxford
Copyright © 1989 Columbia University Press
All rights reserved

Library of Congress
Cataloging-in-Publication Data
Stommel, Henry M., 1920–
An introduction to the Coriolis force /
Henry M. Stommel, Dennis W. Moore.
p.   cm.
Bibliography: p.
Includes index.
ISBN 0-231-06636-8. — ISBN 0-231-06637-6 (pbk.)
1. Coriolis force.   I. Moore, Dennis W.   II. Title.
QC880.4.C65S76   1989
551.5'153—dc19
88-38429
CIP

C 10 9 8 7 6 5 4 3 2 1
P 10 9 8 7 6 5 4 3 2 1

Contents

Acknowledgments

Generally a book of an educational nature originates from a series of lecture notes, worked up over many years. This guarantees that weeds of error will have had plenty of time to sprout and be pruned. Our book was written more spontaneously, as an experiment in reader-computer interaction. We have tried to check every thing carefully, and hope that the weed seeds are few. But they are our responsibility, and we will welcome letters from those who find them. The computer programs offer a fertile ground for new varieties to make themselves shown.

We want to thank Drs. James Luyten and Nelson Hogg for reading over a very early version of the text. Drs. Richard Garvine and K. C. Wong were generous with their comments. The graduate students, Mr. Glen Gawarkiewicz, Mr. Federico Graef-Ziehl and Mr. Kurt Polzin helped us locate some ambiguities in the text and computer displays. Dr. Lew Rothstein of the University of Washington was very helpful with comments on a later version. Mr. John Thomson built the Compton generator. Mr. Richard Rhodes draughted the figures. Mr. Peter Rosti created figures 5.1 and 7.2 on a computer, using Auto-CAD. Dr. Stephen Chiswell and Ms. Sharon Lukas printed out the programs. Mr. Leslie Bialler, Ms. Jennifer Dossin and Mr. Edward Lugenbeel skillfully guided us through the editorial complexities of producing this book.

AN INTRODUCTION
TO THE CORIOLIS FORCE

Introduction

The purpose of this book is to offer a clear physical explanation of the Coriolis force, that somewhat mysterious force invoked by meteorologists and oceanographers to explain the apparent equilibrium of a system of wind patterns or ocean currents in the presence of horizontal pressure gradients, so that the velocity of the fluid lies along isobars. The Coriolis force is conventionally derived by a series of formal manipulations of Newtonian mechanics beginning with a frame of reference supposedly fixed with respect to "the stars" and then transferring it to a reference frame fixed to the rotating earth. All professional meteorologists and oceanographers encounter this mathematical demonstration at an early stage in their education. Clutching the teacher's hand, they are carefully guided across a narrow gangplank over the yawning gap between the resting frame and the uniformly rotating frame. Fearful of looking down into the cold black water between the dock and ship, many are glad, once safely aboard, to accept the idea of a Coriolis force, more or less with a blind faith, confident that it has been derived rigorously. And some prefer never to look over the side again.

So, when they are asked for an explanation by a land-lubber standing on the dock, who has never been carefully guided across the perilous plank, they find themselves singularly unable to explain this curious force. Incomplete explanations of Coriolis force abound in popular books and magazines. The sense of frustration that overcomes those who try to understand explanations of meteorological and oceanographical phenomena can thus be accounted for.

Our little book leads you back and forth across that narrow gangplank. It will force you to look down, and we hope it will cure you of rotation sickness. Our aim is to help you develop a strong physical intuitive understanding of the mechanics of uniformly rotating systems, and to abolish any lingering uncertainty about the Coriolis force.

In order to maintain a sharp focus on the problem of exploring the concept of a Coriolis force we will restrict our discussion to the simple case of moving particles. True, many of the most important applications of the idea of a Coriolis force are to the fluids of the atmosphere and ocean, and therefore really involve hydrodynamics. But the idea of Coriolis force is independent of concepts of pressure, compressibility, and the Eulerian equations of mass conservation and fluid motion. Ideas such as geostrophic motion can be developed partly in terms of individual particle dynamics. Therefore it seems that getting involved in fluid dynamics would be too much of a diversion, and that we might lose focus upon the Coriolis force itself.

Experience has shown that some human minds need specific examples to practice upon and test their understanding. The advent of the microcomputer makes it possible to construct computer assisted interactive programs that work very much like classroom laboratory demonstrations. Therefore the text in this book is supplemented with a series of computer exercises, written in BASIC, that you can copy for your own personal computer, piece by piece. They will provide an interactive tool for experimenting with a variety of problems involving the idea of Coriolis force. The programs are written with unique line numbers so that each can be written on top of the other. As each new program is added, its "Exercise number" can be entered in the handling program. This way you gradually accumulate a complete master program for the whole book. It is a convenient way to avoid having to rewrite various subroutines that are called for in several different cases. The programs were originally written on an NEC-APC III using Microsoft GW-BASIC, Version 2.01, and a color card and monitor. They will run on an IBM-AT with Enhanced Color Display using BASICA. For those who prefer not to make up their own copy of the program a 5 1/4 inch floppy diskette for IBM-PC compatible machines is available through The Market Bookstore, 15 Depot Ave., Falmouth, Mass, 02540. The disk also contains a version of the programs that corrects for the distorted coordinates of the IBM-AT with Professional Graphics Display.

PROLEGOMENON

A big word like prolegomenon bodes of something serious to say: some general remarks at the beginning that will help readers maintain perspective as they go through the detail of the book. These are the main points:

(1) The Newtonian laws of motion hold only in absolute space, resting with respect to a rather vaguely defined inertial system: the "fixed stars." The accelerations of a particle located at x, y, z in a rectilinear coordinate system fixed in a space of absolute rest can be written as $\ddot{x}$, $\ddot{y}$, $\ddot{z}$—double dots indicating two successive time differentia-

tions. If the x, y, z components of force per unit mass (of the particle) are denoted by F_x, F_y, F_z, then by Newton's law

$$\ddot{x} = F_x, \ddot{y} = F_y, \ddot{z} = F_z.$$

The convention is to write the accelerations on the left-hand side of the equation, the forces on the right-hand side.

(2) For many applications it is natural to rewrite these equations in other coordinate systems—for example, polar, cylindrical, and spherical. Such a rewriting is done by a *transformation of coordinates* to another system fixed in the same resting Newtonian absolute space. The results can appear somewhat complicated, involving terms additional to the doubly dotted coordinates. Sometimes the additional terms in the accelerations are transposed to the right-hand side of the equation, leaving only the doubly dotted terms on the left. So the acceleration terms on the right now look like forces. They even acquire names such as "centrifugal force." As convenient as this may be from an intuitive, practical point of view, this transposition of parts of the acceleration to the right-hand side can lead to confusion. It leads to what seem to be new concepts. It is a semantic phenomenon. So remember that the appearance of this type of unreal force does not necessarily involve a rotating coordinate system, moving relative to absolute space. The polar, cylindrical, and spherical coordinates that give them rise are firmly fixed in Newton's safe old system, locked at rest with respect to the "fixed stars."

(3) We happen to live on an earth that is rotating about an axis whose direction is nearly fixed with respect to the stars. It is very useful to employ a system of coordinates that is fixed with respect to the earth (latitude,

longitude, altitude), and this means using a coordinate system rotating uniformly with respect to the stars. When earth-dwellers speak of a particle as being at rest, they mean at rest with respect to the earth itself as a rotating reference frame. When we derive the expressions for acceleration in such a rotating reference frame several new terms appear that make the expressions unfamiliar. Those terms proportional to the relative velocities are called the Coriolis accelerations. Now strictly speaking all of the terms in the accelerations should be written on the left-hand side, but there is an irresistible urge to get the left-hand side into familiar form, so the extra terms are transposed to the right-hand side, where they assume the role of "forces." The Coriolis accelerations become Coriolis forces. The new Coriolis forces arise from the problem's being formulated in a non-Newtonian rotating coordinate system.

(4) Generally there are more terms than just the Coriolis terms in the equations that one wants to shift to the righthand side, in our effort to make the lefthand side look like its counterpart in a nonrotating reference frame. We must face the problem of how to deal with these non-Coriolis forces on the right-hand side. This is where the idea of a *platform of dynamical equilibrium* comes in. By equilibrium we don't mean that in absolute space all forces on particles balance to zero, only that in a particular rotating frame they are arranged to keep the particles in uniform solid rotation about the axis in absolute space. It takes some real forces in absolute space to do this.

We want to introduce some real forces in the absolute reference frame—gravity, reaction of a hard surface on a particle, etc.—that will balance these non-Coriolis terms. Then on the right-hand side the only terms will be the Coriolis forces that are proportional to relative velocity. As a result, with the particle at relative rest anywhere in the rotating system, the Coriolis forces will be zero, and the left-hand side will look exactly like the expression for acceleration that obtains in a nonrotating system. This seeming paradox is attained by the supporting structure of real forces in the absolute system—the platform.

We will construct several platforms upon which particles can rotate about an axis with constant uniform angular velocity in dynamical equilibrium in the following chapters.

(5) You will see that all of this is something of a tortuous semantic trick, justifiable only by its convenience. If we wish to choose a more direct, if often less convenient, way of formulating one of our dynamic problems we can always revert to good old Newtonian space fixed with respect to the stars. Some of our exercises will be computed both ways to illustrate why formulating a problem with reference to the rotating frame is often considerably easier than trying to formulate it in absolute space. On the other hand, the computation of trajectories of free particles moving on surfaces of revolution is often most simply done without introduction of rotating coordinates, by invoking the first integrals: conservation of mechanical energy and angular momentum.

(6) The use of first integrals is closely related to the existence of scalar potentials from which gravitational and centrifugal forces may be derived. To express the force acting upon a particle subject both to gravitation and centrifugal force one simply writes $\mathbf{F} = \nabla\Phi$ where Φ is the sum of the gravitational potential Φ_g and the centrifugal potential $\Phi_\Omega = \Omega^2 r^2/2$, in cylindrical polar coordinates for a system rotating at angular speed Ω about the z axis. The gravitational potential Φ_g of course depends upon the distribution of mass of the attracting body.

A surface of constant potential is called an equipotential surface. A particle rotating with angular velocity Ω

and subject to gravitation experiences a force perpendicular to the equipotential Φ (combined centrifugal and gravitational) on which it lies. If a hard slippery material surface coincides with the equipotential at this point, it exerts a counteracting reaction upon the particle, but no tangential force. We can think of the hard surface as a "platform" upon which the particle can remain at rest in the rotating coordinate system.

These remarks are true for all of the platforms of equilibrium that we consider. However there are important differences between various kinds of platforms. The form of a rotating dish in a laboratory can be fashioned at the will of the experimenter, independent of the structure of the gravitational potential of the earth—which is scarcely affected by the small mass of the dish. On the other hand the figure of the earth itself depends upon its rotation, because the material of the earth is so weak that it behaves like a fluid, and is free to assume its own appropriate shape for rotating equilibrium. In this case we prescribe neither the shape of the earth a priori, nor the form of the gravitational potential. We must know the unknown figure in order to compute the gravitational potential—so the figure of the earth becomes an implicit problem. This is the problem for which Maclaurin found his beautiful solution. He was able to show for a fluid earth of uniform density, rotating with a certain angular velocity [within limits], there are two ellipsoidal surfaces bounding the mass which can coincide with a constant value of the combined gravitational potential [for the ellipsoidal mass] and centrifugal potential.

(7) In meteorological and oceanographical applications the concept of a platform of equilibrium is generalized, somewhat beyond that of a simple particle sliding upon a slippery hard surface. A thin layer of fluid is envisaged, covering the equilibrium surface of the ellipsoid, and at

rest with respect to the rotating coordinate system. The hydrostatic pressure field in the fluid layer coincides with the equilibrium surfaces of the combined potentials, balancing the combined force exactly. In this way the pressure field plays much the same role as the reactive force of the hard surface plays in our simpler platforms.

(8) A prolegomenon ought to be reread several times as one goes through the text and examples. It ought to become clearer each time, until it finally embraces the meaning of the whole book.

CHAPTER I

Real and apparent force

1.1 REAL FORCE

It helps at first to think about motions in a horizontal plane, the x,y plane that we envisage as being at rest with respect to "the stars." We could think of it as a very large horizontal frictionless table top, so that the only effect of gravity would be to hold the particles against the table with a gravity force perpendicularly downward toward the table top, and that this is balanced in the vertical, z, by a reacting force from the table's surface. Therefore particles don't accelerate in the vertical: gravity and reactive table force just balance. This leaves us with the horizontal forces and accelerations due to them. Of course we really don't need gravity to hold the particles to the table top, because horizontal forces wouldn't move them away from there anyway, but the presence of gravity does make things feel concrete. It also helps to orient our hypothetical observer: he knows which way his feet are pointing relative to the table top. In the absence of horizontal forces a particle at point x,y is free to move horizontally with any constant horizontal components of velocity: $\dot{x},\dot{y}$ [we find it convenient to denote time derivatives by dot notation]. If there is a horizontal force $\mathbf{F}$ per unit mass applied, with components F_x and F_y, then the particles will accelerate according to the Newtonian axiom by the equations

$$\ddot{x} = F_x, \ \ddot{y} = F_y. \tag{1.1}$$

These are the dynamical equations in a resting reference frame. In the absence of a force, $F_x = F_y = 0$, then $\ddot{x} = \ddot{y} = 0$ and the particle has a constant velocity $\dot{x} = \dot{x}_i$, $\dot{y} = \dot{y}_i$ where the subscript i denotes the word "initial." The quantities $\dot{x}_i$ and $\dot{y}_i$ are therefore the velocities at time $t = 0$ [the "initial" time]. If, during the course of time the forces turn on, with some time dependence of the general form $F_x = F_x(t)$, $F_y = F_y(t)$, then by integration of equation (1.1) we obtain for the velocities

$$\dot{x}(t) = \dot{x}_i + \int_0^t F_x(\tau)\, d\tau$$

$$\dot{y}(t) = \dot{y}_i + \int_0^t F_y(\tau)\, d\tau,$$

and with one further integration we obtain

$$x(t) = x_i + \int_0^t \dot{x}(\tau)\, d\tau$$

$$y(t) = y_i + \int_0^t \dot{y}(\tau)\, d\tau.$$

Figure 1.1 shows the trajectory of a particle by a series of dots that indicate the position at uniform increments of time. For the first ten time increments the applied forces are zero so the particle moves steadily with constant initial velocity $\dot{x}_i = 0$, $\dot{y}_i = \text{const}$. At time increment 10 the force $F_x = \text{const}$, $F_y = 0$ turns on so the particle begins to accelerate in the positive x-direction, as shown in the fig-

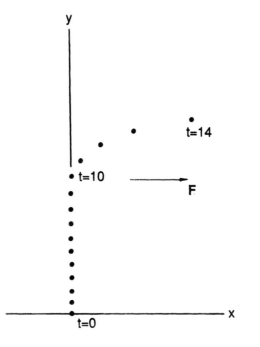

Figure 1.1. The successive positions of a particle sliding on the frictionless *x, y* plane are shown by dots, beginning at the origin. After ten time intervals with no force and a uniform constant velocity in the *y* direction, a force in the *x* direction is turned on, as indicated by the arrow. In addition to the constant drift in the *y* direction the particle begins to accelerate in the *x* direction. If you were to observe a trajectory of this kind you would probably infer the onset of the force.

ure. We have a strong visual impression of the sudden operation of the force acting at right angles to the original direction of movement before the turning on of the force. Drifting uniformly in the positive y direction the particle is literally "blown" toward positive x beginning at the tenth time increment.

1.2 APPARENT FORCE

We now want to compare the visual image of what we have just seen in figure 1.1 with what we observe when we are moving with a uniformly rotating reference frame, when there is in fact no force turned on. The observer intuits that a force is acting even when there is no force. The impression is so strong that it is actually useful, and leads to the invention of the class of nonexistent, apparent, virtual, fake or adventitious forces called "centrifugal force" and "Coriolis force."

Suppose that there is a system of rectilinear coordinates x', y' rotating uniformly at the rate Ω about the origin of the x, y system [the reference frame at rest with respect to the "stars"]. At time $t = 0$ the two sets of coordinates coincide. At any time t the two sets of coordinates are related by

$$x = x' \cos(\Omega t) - y' \sin(\Omega t)$$
$$y = x' \sin(\Omega t) + y' \cos(\Omega t), \tag{1.2}$$

or*

$$x' = x \cos \Omega t + y \sin \Omega t$$
$$y' = -x \sin \Omega t + y \cos \Omega t. \tag{1.3}$$

*From now on we will omit the parentheses around the argument of trigonometric functions when it is a simple product.

Now let us pose a very simple problem. We imagine that there are no forces at all, that a particle moves through the x,y plane with uniform velocity $\dot{x}$ = constant, $\dot{y} = 0$, passing through the y axis at $t = 0$. This straight trajectory in the x,y plane appears as a complicated curve in the x',y' plane (figure 1.2). There is a strong visual impression of the particle's being accelerated by a mysterious force associated with the origin in some complicated way. The analytical expressions for these apparent forces can be obtained by differentiating the equations (1.3) twice with respect to time to get the $\ddot{x}'$ and $\ddot{y}'$ in the rotating reference frame. Because they are not zero, their values on the righthand side appear to be forces, "apparent" forces. Noting that in the special case we displayed that the solution in the absolute frame is $x = \dot{x}_i t$ and $y = y_i$, the first differentiation gives

$$\dot{x}' = \dot{x}_i \cos\Omega t - \Omega(x\sin\Omega t - y\cos\Omega t)$$

$$\dot{y}' = -\dot{x}_i \sin\Omega t - \Omega(x\cos\Omega t + y\sin\Omega t),$$

and the second differentiation gives

$$\ddot{x}' = -2\Omega\dot{x}_i\sin\Omega t - \Omega^2(\dot{x}_i t\cos\Omega t + y_i\sin\Omega t)$$

$$\ddot{y}' = -2\Omega\dot{x}_i\cos\Omega t - \Omega^2(-\dot{x}_i t\sin\Omega t + y_i\cos\Omega t).$$

Now just look at the righthand side with those odd-looking combinations of the quantities Ω, $\dot{x}_i$, y_i and t, all of which we know. They look like "forces" balancing accelerations $\ddot{x}'$, $\ddot{y}'$ in the rotating system.

Thus there are apparently several kinds of force acting, rotating with frequency Ω and out of phase. The first depends upon the product of the angular velocity of rotation and the absolute velocity. The second has a factor Ω^2 and increases linearly with time. The third depends upon the factor Ω^2 and the absolute constant y_i coordinate of the

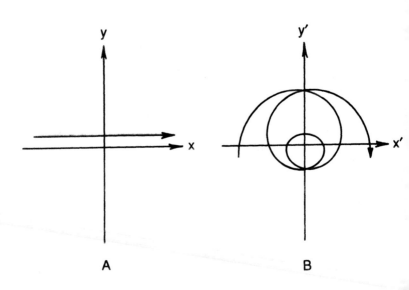

Figure 1.2. In the lefthand panel you see the trajectory of a particle drifting under no force in the positive x direction with uniform x velocity. The trajectory is a nice straight line. If you were to observe this same particle from an x', y' plane that rotates uniformly counterclockwise, then the trajectory would appear as in the righthand panel. The gyrating trajectory would suggest to you a rather complicated force acting on the particle, but there is no force. You just think that there might be, unless of course you happen to look up at the sky and see the "fixed" stars going around.

particle in absolute space. These are the "fake" forces that one intuits to explain the gyrations of the particle observed in Figure 1.2.

We note in passing that the equations for $\dot{x}'$ and $\dot{y}'$ above can also be written in the form

$$\dot{x}' = \dot{x}_i\cos\Omega t + \Omega y', \qquad \dot{y}' = -\dot{x}_i\sin\Omega t - \Omega x'.$$

These expressions for $\dot{x}'$ and $\dot{y}'$ are implicit rather than explicit in time, since they define the velocity in terms of time and the present position of the particle in the coordinate system that is rotating. The previous expressions were variable only in time. Implicit expressions may seem confusing but are often more compact to write, and can make the physical interpretation more direct. For example, if we take the initial velocity $\dot{x}_i = 0$ so that our particle is at rest at $(0, y_i)$ in the absolute inertial space, then our equations in the rotating system are

$$\dot{x}' = \Omega y', \qquad \dot{y}' = -\Omega x'.$$

These expressions describe the motion of a particle rotating with angular velocity $-\Omega$ [i.e. clockwise for $\Omega > 0$] around the origin. The second derivative gives

$$\ddot{x}' = \Omega\dot{y}' = -\Omega^2 x', \qquad \ddot{y}' = -\Omega\dot{x}' = -\Omega^2 y',$$

indicative of a radially inward directed "force."

In the example of figure 1.2 the angular velocity of the rotation is not fixed to anything particularly physical, so we cannot attach any particular physical significance to the rotating frame. When we find a physical context within which to make such an identification of Ω we will have constructed in our minds a meaningful rotating *platform*. There are, of course, no new forces added to a physical system in absolute space just because we prefer for some reason or other to refer it to a rotating system. However, we earth-dwellers do live on a rotating system, and ordinarily find it convenient to refer our physics to it—at the cost, though, of introducing exotic pseudo-forces: virtual, fake, apparent, adventitious—however we choose to denote them.

EXERCISES

The exercises are in the form of computer programs with which one can experiment, as in a laboratory. They are written in BASIC for IBM-PC compatible machines. It is convenient to present the programs for the various exercises as separate programs. The exercises are numbered in the format *M-N*, and the corresponding programs begin with line numbers in full thousands. This leaves room at small line numbers for a general handling program that permits rapid access to the programs, and acts as a table of contents. It also leaves room at the end of the program, lines 60000 +, for some useful subroutines that draw arrows etc. Some of these subroutines are called up by some of the programs. If your machine has a limited internal memory then it may not be able to load the combined programs all at once. In this case you will dispense with the handling program and use each exercise separately. If you do, then remember to include the subroutines for exercises that need them, by merging the subroutines with the exercise in question.

The main line numbers referred to are by tens. For typographic reasons intermediate steps are in units. You can type up each program separately and run it without having to have all of them on the disk. You can build up the whole set piece by piece.

To start building the handling program, one writes the lines 1−300 on the disk.

```
  1 GO TO 55
 10 ' Program name :CORIOLIS
 11 ' Edited April 1988
 50 ' handling program    ********************
 55   SCREEN 1: COLOR 0,2: KEY OFF: CLS :
 60   LOCATE 5,10
100 INPUT "EXERCISE NUMBER";E$
111 IF E$ = "1-1" THEN GOTO 5000
112 IF E$ = "1-2" THEN GOTO 6000
121 IF E$ = "2-1" THEN GOTO 10000
122 IF E$ = "2-2" THEN GOTO 11000
123 IF E$ = "2-3" THEN GOTO 13000
124 IF E$ = "2-4" THEN GOTO 14000
131 IF E$ = "3-1" THEN GOTO 15000
132 IF E$ = "3-2" THEN GOTO 16000
133 IF E$ = "3-3" THEN GOTO 17000
134 IF E$ = "3-4" THEN GOTO 18000
141 IF E$ = "4-1" THEN GOTO 20000
151 IF E$ = "5-1" THEN GOTO 25000
152 IF E$ = "5-2" THEN GOTO 26000
153 IF E$ = "5-3" THEN GOTO 27000
154 IF E$ = "5-4" THEN GOTO 28000
155 IF E$ = "5-5" THEN GOTO 29000
156 IF E$ = "5-6" THEN GOTO 24000
157 IF E$ = "5-7" THEN GOTO 24500
165 IF E$ = "6-1" THEN GOTO 34000
166 IF E$ = "6-2" THEN GOTO 33000
171 IF E$ = "7-1" THEN GOTO 30000
172 IF E$ = "7-2" THEN GOTO 31000
173 IF E$ = "7-3" THEN GOTO 32000
181 IF E$ = "8-1" THEN GOTO 35000
182 IF E$ = "8-2" THEN GOTO 36000
183 IF E$ = "8-3" THEN GOTO 37000
184 IF E$ = "8-4" THEN GOTO 38000
191 IF E$ = "9-1" THEN GOTO 47000
```

```
192 IF E$ = "9-2" THEN GOTO 48000
193 IF E$ = "9-3" THEN GOTO 49000
194 IF E$ = "9-4" THEN GOTO 54000
195 IF E$ = "A-1" THEN GOTO 55000
196 IF E$ = "a-1" THEN GOTO 55000
300 GOTO 55
```

CONVENTIONS ABOUT NOTATION

We will adhere to a fixed notation for variables used in the exercises throughout this book. The variables will be denoted in basic as follows

$$X = x \quad Y = y \quad Z = z \quad R = r \quad P = \phi \quad TH = \vartheta$$

$$MX = x' \quad MY = y' \qquad\qquad MP = \phi'$$

$$XD = \dot{x} \quad XDD = \ddot{x} \quad \&c.$$

We denote the primed quantities (generally denoting reference to moving reference frames) by the prefix M meaning prime "mark" or "moving" reference frame. The suffix D is used to denote the time differentiation "dot." A prefix D usually denotes a differential, such as $DT = dt$.

Exercise 1-1 [Lines 5000−5160 and subroutines 1 and 2] *Particle accelerating under suddenly turned-on force.* A particle starts at the origin [5030] with a randomly chosen *XD* and *YD*. The integration of the dynamical equations [5050−5070] is carried out for the first ten time steps with the force components *FX* and *FY* set to zero, but when [5100−5110] time *T* is equal to 10*$*dt$ the force components *FX* and *FY* are set equal to the constants *FX0, FY0* so the particle begins to accelerate in the direction of the force. The force is displayed as an arrow with tail at *AX,AY* [5120], length components *LX,LY*, angle of the arrow head *AL*, and length of the head *LH*. The arrow is drawn by Subroutine 1 [60000−60080]. When the particle goes off screen, the screen is cleared [5150] the amplitudes of the force components *FX0, FY0* are chosen randomly again [5040] and the whole process repeated. One hopes that this display gives a vivid impression of the effect of a fixed force suddenly turned on in an inertial system (a system of rectilinear coordinates fixed with respect to "the stars").

```
5000 ' Exercise 1-1 ••••••••••••••••••••••••••
5005 SCREEN 1:     KEY OFF:CLS:COLOR 0,2
5010 DT= 1 : FAC = 10:FFF=10: FFC = 10
5020 CLS
5030 X = 0: Y=0: XD= FFC*(RND-1/2)
5031 YD=FFC*(RND-1/2):T=0: CA = 3
5040 FX0= FFF*(RND-1/2):FY0=FFF*(RND-1/2)
5050  XDD=FX: YDD=FY
5060 XD=XDD*DT+XD:YD=YDD*DT+YD
5070 X = XD*DT+X: Y=YD*DT+Y
5080 PSET (150+X, 100-Y),2
5090 T = T+DT: FX=FX0:FY=FY0
5100 IF T<(11*DT) THEN FX=0
5110 IF T<(11*DT) THEN FY = 0
5120 AX=150:AY=80:LX=FAC*FX
5121 LY = FAC*FY : AL=.4 : LH = 4
5130 IF T<(11*DT) THEN LH = 0 ELSE LH=4
5140 GOSUB 60000
5150 IF X^2+Y^2>10000 THEN CLS:    GOTO 5020
5160 GOTO 5050
```

```
60000 ' Subroutine 1 ••••••••••••••••••••••••••
60005 PI = 3.14159:' arrow subroutine
60010 LINE(AX,200-AY)-(AX+LX,200-AY-LY),CA
60020 IF LX = 0 THEN TH=SGN(LY)*PI/2
60030 IF LX = 0 THEN GOTO 60060
60040 TH = ATN(LY/LX)
60050 IF LX>0 THEN TH = TH ELSE TH = TH + PI
60060 XZ=AX+LX-LH*COS(TH+AL)
60061 YZ=200-(AY+LY-LH*SIN(TH+AL))
60062 LINE(AX+LX,200-(AY+LY))-(XZ,YZ),CA
60070 XZ=AX+LX-LH*COS(TH-AL)
60071 YZ=200-(AY+LY-LH*SIN(TH-AL))
60072 LINE(AX+LX,200-(AY+LY))-(XZ,YZ),CA
60080 RETURN
```

```
61000 'Subroutine 2;Draw star ***************
61010 PSET(XLS,YLS),0:GOSUB 61040
61020 PSET(XS,YS),3:GOSUB 61040
61030 RETURN
61040 FOR D=0 TO 360 STEP 72
61050 DRAW "ta=d;u3
61060 NEXT D
61070 RETURN
```

Exercise 1-2 [Lines 6000−6250] *Motion without forces observed from fixed and rotating frames of reference.* The motion is computed by stepping X at a constant rate [6140] from an initial position X, Y [6040]. There is room to experiment with this program because one can change the initial positions in [6040], or the steady steps in X and Y [6140]. There are a fair number of graphical statements in this program that we can ignore talking about. After we get the new X and Y we compute the angle P [6150] and R [6050]. Also we step up the angle of rotation TH of the moving coordinates [6130] and compute the relative polar coordinate angle MP [6160]. Then we compute the rectilinear coordinates MX and MY in the moving frame [6190−6200] and after a delay to slow the display down [6210], plot them [6220−6230] in the two reference frames. The great deflection of the particle in the moving reference frame gives a strong impression of a force acting, but as we know there is no force: the particle is moving with constant velocity in the absolute reference frame.

```
6000 ' exercise 1-2 ••••••••••••••••••••••••••
6010 SCREEN 1: COLOR 0,2: KEY OFF: CLS
6020  R = SQR(X^2+Y^2)
6030 PI = 3.14159265#:TH=PI/4
6040  X=-70:Y=10
6050  R = SQR(X^2+Y^2)
6060 LINE(0,100)-(149,100),1
6070 LINE(151,100)-(300,100),1
6080 LINE(75,25)-(75,175 ),1
6090 LINE(225,25 )-(225,175),1
6100 LOCATE 12,37:PRINT "mx"
6101 LOCATE 12,18:PRINT"x"
6110 LOCATE 3,28: PRINT "my"
6111 LOCATE 3,10: PRINT "y"
6120 LOCATE 22,6:PRINT"absolute"
6121 LOCATE 22,24: PRINT "relative"
6130 TH = TH+.1
6140 X = X+.99
6150 P  = ATN( Y/X):IF X<0 THEN P =P +PI
6160 MP =P -TH
6170 A = R*COS(PHP)
6180 AX = A*COS(TH):AY=A*SIN(TH)
6190 MX = X*COS(TH)+Y*SIN(TH)
6200 MY = -X*SIN(TH)+Y*COS(TH)
6210 FOR I = 1 TO 50: ZZ=100^2: NEXT
6220 PSET(75+X,100-Y),3
6230 PSET(225+MX,100-MY),3
6240 LOCATE 1,1:THH=TH*180/PI
6241 PRINT USING "theta=#### degrees   ";THH
6250 GOTO 6130
```

CHAPTER II

Velocity and acceleration in plane polar coordinates

2.1 TRANSFORMATION OF COORDINATES

We are now going to consider how to think of velocity and acceleration in coordinates other than rectilinear ones x, y. The first new set of coordinates is going to be plane polar coordinates. There is going to be an angle introduced, and we will also introduce a radius measured from a central fixed point at the origin. You should be very clear in your mind not to confuse this idea of a plane polar coordinate system fixed in inertial space with the idea of rotation of coordinates. This chapter is entirely tied to one particular reference frame, fixed in inertial space—so don't get mixed up now, or later when we introduce rotating axes.

It is often convenient to represent the position P of a particle by r, its distance from the origin, and an angle ϕ measured counterclockwise from the x-axis to the radius OP (figure 2.1). Thus we have two ways of representing the position, and from simple geometry they are related by the equations:

$$x = r\cos\phi, \qquad y = r\sin\phi \tag{2.1}$$

$$r = \sqrt{x^2 + y^2} \quad \text{and} \quad \phi = \tan^{-1}\left(\frac{y}{x}\right).$$

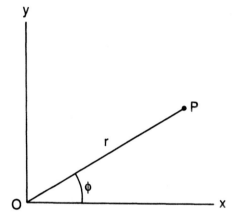

Figure 2.1. The position of a point P can be specified in the x, y plane by rectilinear coordinates x and y. It can also be specified by the radial distance r and the angle ϕ that the radius OP makes with the x axis. This alternate way of specifying the position of point P is called polar coordinates.

2.2 VELOCITY AND ACCELERATION

We can imagine, for example, that x and y are some single valued functions of time t. We can therefore compute a trajectory of the particle on the x, y plane by varying t. If we calculate the time derivatives of $x(t)$ and $y(t)$ we then have expressions for the velocity components $\dot{x}(t)$, $\dot{y}(t)$. If we once again calculate the time derivatives of these we have the rectilinear accelerations $\ddot{x}(t)$, $\ddot{y}(t)$.

Let us denote the directed line segment from O to P by the vector **P**.

The velocity can be visualized graphically by choosing two successive times t and $t + dt$, a short time interval dt apart, and computing the positions at these two times as shown in figure 2.2a. The change in position, $d\mathbf{P}$, measured between the two positions, divided by the time interval dt, is a measure of the *velocity* **v**. We can plot this velocity vector on another diagram, figure 2.2b, on the $\dot{x}$, $\dot{y}$ plane. It executes, with time, a curve called the hodograph. It has a length V and a direction ω. The two directions ϕ and ω are generally not the same. We can now do the same thing with the hodograph that we did in position space. We draw two locations of the velocity on the hodograph at times t and $t + dt$. The distance $d\mathbf{v}$ between these two points, when divided by the time interval dt is the *acceleration*. We can draw the acceleration on the $\ddot{x}$, $\ddot{y}$ plane, as in figure 2.1c. We now have another magnitude for the acceleration, A, and another direction α.

The x, y components of velocity $\dot{x}$, $\dot{y}$ are computed as functions of the polar quantities by differentiating equations (2.1) with respect to time:

$$\dot{x} = \dot{r}\cos\phi - r\dot{\phi}\sin\phi$$
$$\dot{y} = \dot{r}\sin\phi + r\dot{\phi}\cos\phi, \qquad (2.2)$$

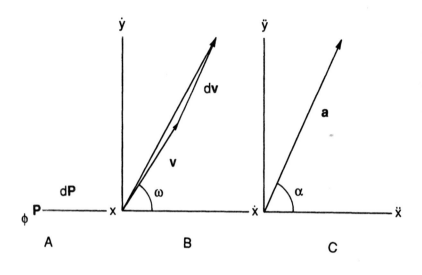

Figure 2.2. The position vector of point *P* in polar coordinates may change over a time interval *dt* by the amount *d***p** (lefthand panel). When you divide *d***p** by *dt* you get the velocity **v** of the particle. This directed quantity (vector) can be drawn in the *ẋ, ẏ* plane as the vector **v** with magnitude *V*. It has an angle *ω* that is different from the *φ* of the position vector. With time the point of the **v** vector describes a curve in the *ẋ, ẏ* space, called a hodograph (middle panel). Over a time interval *dt* the vector **v** increases by an amount *d***v**. When divided by *dt* this new arrow defines an acceleration **a**. The vector **a** has another angle *α* in the *ẍ, ÿ* space. With time it can also change, but Newtonian dynamics is formulated in terms of accelerations, so we don't need further names for the obvious endless regression of constructions that is implied.

and the x, y components of acceleration, $\ddot{x}$, $\ddot{y}$ are obtained by one more time differentiation:

$$\ddot{x} = \ddot{r}\cos\phi - 2\dot{r}\dot{\phi}\sin\phi - r\dot{\phi}^2\cos\phi - r\ddot{\phi}\sin\phi \quad (2.3)$$

and

$$\ddot{y} = \ddot{r}\sin\phi + 2\dot{r}\dot{\phi}\cos\phi - r\dot{\phi}^2\sin\phi + r\ddot{\phi}\cos\phi.$$

Notice that we have three generally different angles to contend with: ϕ, ω and α. We want to do a little projective geometry to get relations between the x, y and r, ϕ components of the velocity and acceleration. Figure 2.3A is in the x, y plane [note that this is different from the figures 2.2B and C]. We draw the r and ϕ coordinates as well as the x, y ones, both to the same point. Now we draw the velocity at this point in figure 2.3A and figure 2.3B, the acceleration in figure 2.4A and figure 2.4B. We also draw the components. The components of the vectors **v** and **a** in the rectilinear axes are written in the dotted form: $\dot{x}$, $\dot{y}$, and $\ddot{x}$, $\ddot{y}$. The polar projections on the radius, r, and angle ϕ are denoted by v_r, v_ϕ and a_r, a_ϕ, for velocity and acceleration respectively.

From figure 2.3 it is clear that the polar coordinates components of velocity are

$$\begin{aligned} v_r &= V\cos(\omega - \phi) \equiv V(\cos\omega\cos\phi + \sin\omega\sin\phi) \\ v_\phi &= V\sin(\omega - \phi) \equiv V(\sin\omega\cos\phi - \cos\omega\sin\phi) \end{aligned} \quad (2.4)$$

but since the velocity components in rectilinear coordinates are

$$\dot{x} = V\cos\omega, \qquad \dot{y} = V\sin\omega \quad (2.5)$$

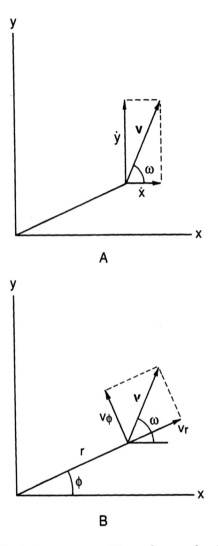

Figure 2.3. A given vector **V** can be resolved into components in the x and y directions or the r and ϕ directions. The two panels show the relations between them, for the case of a velocity.

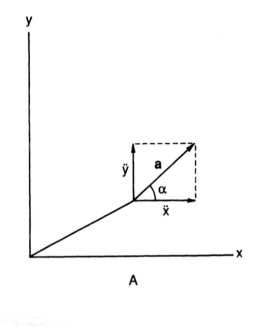

A

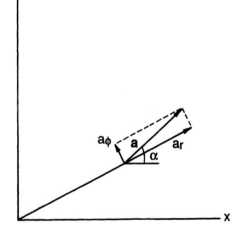

B

we can write the polar components of velocity in terms of the components of velocity as

$$v_r = \dot{x}\cos\phi + \dot{y}\sin\phi \qquad (2.6)$$

and

$$v_\phi = -\dot{x}\sin\phi + \dot{y}\cos\phi.$$

We can now eliminate the $\dot{x}$ and $\dot{y}$ by using expressions (2.2) to obtain

$$v_r = (\dot{r}\cos\phi - r\dot{\phi}\sin\phi)\cos\phi + (\dot{r}\sin\phi + r\dot{\phi}\cos\phi)\sin\phi$$

$$v_\phi = -(\dot{r}\cos\phi - r\dot{\phi}\sin\phi)\sin\phi + (\dot{r}\sin\phi + r\dot{\phi}\cos\phi)\cos\phi.$$

$$(2.7)$$

Noting the cancellations and trigonometric identities we finally have the polar form of the components of velocity in terms of the polar coordinates and their derivatives as follows

$$v_r = \dot{r}, \qquad v_\phi = r\dot{\phi}. \qquad (2.8)$$

It comes as no surprise to find that the radial component of velocity is $\dot{r}$. However one might ponder a bit over why the angular component of velocity is $r\dot{\phi}$, rather than $\dot{r\phi}$, for example.

Figure 2.4 [*opposite page*]. Similar to figure 2.3, we can draw the components of an acceleration **a** in either the x and y directions or the r and ϕ directions.

When we proceed to finding the components of acceleration in polar form we go through the construction of projections exactly as before (figure 2.4) except that now the expressions for eliminating x and y (equations 2.3) are more involved. Beginning with

$$a_r = A\cos(\alpha - \phi) \equiv A(\cos\alpha\cos\phi + \sin\alpha\sin\phi)$$
$$a_\phi = A\sin(\alpha - \phi) \equiv A(\sin\alpha\cos\phi - \cos\alpha\sin\phi), \qquad (2.9)$$

and noting that

$$\ddot{x} = A\cos\alpha, \qquad \ddot{y} = A\sin\alpha, \qquad (2.10)$$

we write

$$a_r = \ddot{x}\cos\phi + \ddot{y}\sin\phi, \qquad (2.11)$$

and

$$a_\phi = -\ddot{x}\sin\phi + \ddot{y}\cos\phi.$$

Eliminating the $\ddot{x}$ and $\ddot{y}$ by use of equations (2.3) we obtain the cumbersome form

$$a_r = [(\ddot{r} - r\dot{\phi}^2)\cos\phi - (r\ddot{\phi} + 2\dot{r}\dot{\phi})\sin\phi]\cos\phi$$
$$\qquad + [(\ddot{r} - r\dot{\phi}^2)\sin\phi + (r\ddot{\phi} + 2\dot{r}\dot{\phi})\cos\phi]\sin\phi \qquad (2.12)$$

and

$$a_\phi = -[(\ddot{r} - r\dot{\phi}^2)\cos\phi - (r\ddot{\phi} + 2\dot{r}\dot{\phi})\sin\phi]\sin\phi$$
$$\qquad + [(\ddot{r} - r\dot{\phi}^2)\sin\phi + (r\ddot{\phi} + 2\dot{r}\dot{\phi})\cos\phi]\cos\phi.$$

Noting the cancellations and trigonometric identities, these reduce to

$$a_r = \ddot{r} - r\dot{\phi}^2, \tag{2.13}$$

and

$$a_\phi = r\ddot{\phi} + 2\dot{r}\dot{\phi}.$$

We immediately see that the radial acceleration is not just a simple $\ddot{r}$ as we might have expected by induction from our determination of the radial velocity. Also the angular acceleration is not simply $r\ddot{\phi}$ or $r\dot{\phi}$ as we might have guessed from the angular velocity. So, when we want to write down Newton's equations of motion in polar form, where the components of the force are also written in polar form, F_r and F_ϕ, they must be expressed as

$$\ddot{r} - r\dot{\phi}^2 = F_r, \tag{2.14a}$$

and

$$r\ddot{\phi} + 2\dot{r}\dot{\phi} = F_\phi, \tag{2.14b}$$

where the left hand sides are the acceleration components in polar form and the right hand sides are the force components.

An alternative presentation using vector notation is also possible. Let $\mathbf{i}$ and $\mathbf{j}$ be unit vectors in the x and y directions, so that the position of the particle can be written as $\mathbf{P} = x\mathbf{i} + y\mathbf{j}$. In polar coordinates (r, ϕ) this becomes $\mathbf{P} = r\cos\phi\mathbf{i} + r\sin\phi\mathbf{j} = r\boldsymbol{\rho}$ where $\boldsymbol{\rho} = \cos\phi\mathbf{i} + \sin\phi\mathbf{j}$ is a unit vector pointing away from the origin at the point P. A unit vector at right angles to $\boldsymbol{\rho}$ is

$$\boldsymbol{\sigma} = -\sin\phi\mathbf{i} + \cos\phi\mathbf{j}.$$

This vector points in the direction of increasing ϕ from the point P. Now whereas $\mathbf{i}$ and $\mathbf{j}$ are fixed in space, $\boldsymbol{\rho}$ and $\boldsymbol{\sigma}$

are not, and as ϕ varies with time we have $\dot{\boldsymbol{\rho}} = \dot{\phi}\boldsymbol{\sigma}$ and $\dot{\boldsymbol{\sigma}} = -\dot{\phi}\boldsymbol{\rho}$. From these relations we may easily derive the velocity and acceleration in polar coordinates. Since $\mathbf{P} = r\boldsymbol{\rho}$, we have

$$\mathbf{V} = \dot{\mathbf{P}} = \dot{r}\boldsymbol{\rho} + r\dot{\boldsymbol{\rho}} = \dot{r}\boldsymbol{\rho} + r\dot{\phi}\boldsymbol{\sigma}$$

and

$$\mathbf{a} = \ddot{\mathbf{P}} = \ddot{r}\boldsymbol{\rho} + \dot{r}\dot{\boldsymbol{\rho}} + (\dot{r}\dot{\phi} + r\ddot{\phi})\boldsymbol{\sigma} + r\dot{\phi}\dot{\boldsymbol{\sigma}}$$
$$= (\ddot{r} - r\dot{\phi}^2)\boldsymbol{\rho} + (r\ddot{\phi} + 2\dot{r}\dot{\phi})\boldsymbol{\sigma}.$$

This immediately gives the components of acceleration in polar coordinates, and if the force per unit mass on the particle is written as $\mathbf{F} = F_r\boldsymbol{\rho} + F_\phi\boldsymbol{\sigma}$ we obtain

$$\ddot{r} - r\dot{\phi}^2 = F_r, \qquad r\ddot{\phi} + 2\dot{r}\dot{\phi} = F_\phi$$

as before.

Remember once again that all this has nothing to do with rotating coordinate systems. We are in a polar coordinate system that is at rest with respect to the stars.

Sometimes equation (2.14a) is written with one of the acceleration terms on the righthand side

$$\ddot{r} = r\dot{\phi}^2 + F_r.$$

The term $r\dot{\phi}^2$ then looks like a force, and it actually has a name: "the centrifugal force" [per unit mass]. It is always positive and directed away from the origin. But it is really not a force at all, and so if we want to make use of it in a formal sense, then we could call it a virtual, fake, adventitious force. If we set the true radial component of force to zero ($F_r = 0$) then we cannot have $\ddot{r} = 0$ unless simultaneously we have $\dot{\phi} = 0$. Or again, consider the simple case of free motion where $x = \text{const} \cdot t$, $y = y_0$ in rectilinear coordinates. There is no force acting and the particle moves uniformly in the positive x direction. The coordinates r and ϕ change in time, as do the single and double dotted $\ddot{r}$ and $\ddot{\phi}$ as well. We can now see that $\ddot{r}$ and $r\ddot{\phi}$ cannot possibly be the full expressions for accelerations in the polar coordinates, which must in this case be zero. Sometimes a contradiction of this kind helps illustrate why matters have to be more complicated than one would hope.

The equation (2.14b) also has a conventional interpretation, but a more comfortable one. Multiplying by r, the two terms on the left can be combined, we obtain

$$\overline{r^2\dot{\phi}} = rF_\phi. \tag{2.15}$$

The expression $r^2\dot{\phi}$ is called the angular momentum [per unit mass] of the particle around the origin. The equation states that the time derivative of the angular momentum is equal to the torque rF_ϕ of the force F_ϕ about the origin of the coordinate system.

In this chapter we have faced the fact that there is something of a crisis in intuition that arises from the introduction of the polar coordinate system, even in a nonrotating system or reference frame. When we first use rectilinear coordinates x, y to understand the dynamics of a particle, we commit our minds to the simple expressions $\ddot{x} = F_x$, $\ddot{y} = F_y$. We think of the accelerations as time rate of change [per unit mass] of the linear momentum $\dot{x}$, $\dot{y}$. Then we express the same situation in polar coordinates and are distressed not to find an apparently similar form, such as $\ddot{r} = F_r$, $r\ddot{\phi} = F_\phi$. So we invent some new concepts that partly restore the wanted form. In the case of the radial component of the acceleration we move the $r\dot{\phi}^2$ term to the right hand side and call it a "centrifugal force." In the case of the second equation we formulate a new idea, involving angular momentum and torque. These reformulations of the ideas of dynamics are, in a sense, a mere invention of new names. They are useful, but can be confusing.

Here is a slightly different derivation of the velocity and acceleration components in plane polar form that we like a little better than the one above.

Let us imagine any directed quantity **V** that is given in x and y components V_x and V_y. We want to express **V** in terms of polar coordinates. Consider that it is at point P in figure 2.5, whose coordinates are given in rectilinear coordinates by x, y and in polar coordinates by r, ϕ. We could draw an arrow to show this quantity at P but we won't. Instead we'll draw the x and y components. Now remember that the lengths of these arrows occupy no space in the plane; only the tails of the arrows do. If we draw a line through P perpendicular to r then we have the direction of the ϕ component of this same quantity, and if we extend the radius, we have the direction of the r component of the same quantity. Our job is to project each of the components V_x and V_y onto these new lines. In general there will be a contribution from each. So V_ϕ will be the sum of the two projections, from the rectilinear components on the perpendicular line PP', and V_r will be the sum of the two projections from the rectilinear components on the extension of line OP. One of them is clearly negative, so watch out for that. The formulas expressing these sums are

$$V_r = V_x\cos\phi + V_y\sin\phi, \qquad (2.16)$$

and

$$V_\phi = -V_x\sin\phi + V_y\cos\phi.$$

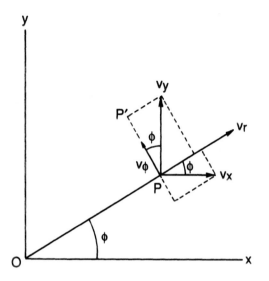

Figure 2.5. This diagram shows a direct way of projecting the *x* and *y* components of a vector on to the extended radius *r* and the line perpendicular to it through the tails of the arrows to obtain the components of the same vector in *r, φ* coordinates. As you can see both the V_x and V_y components project upon the extended radius, so you have to add the two projections up to get the V_r; similarly there are two projections to add (algebraically) to get the $V_φ$. Here **V** stands for the word *vector,* not necessarily *velocity.*

Suppose we now consider that **V** represents velocity. Then the x and y components of velocity are $\dot{x}$, $\dot{y}$. From the equations (2.2) we can write these in terms of the polar coordinates and their time derivatives, so we find

$$v_r = \dot{r}, \qquad v_\phi = r\dot{\phi}. \qquad (2.17)$$

Now we can decide that our quantity **V** is really meant to be the acceleration, so we had better now call it **A**. Our x, y components of **A** are now A_x, A_y. We have in equations (2.3) suitable expressions for these in terms of the polar coordinates and their time derivatives, but they are more involved. We again obtain

$$A_r = \ddot{r} - r\dot{\phi}^2, \qquad (2.18)$$

and

$$A_\phi = r\ddot{\phi} + 2\dot{r}\dot{\phi}.$$

The advantage of this derivation is that one doesn't have to think explicitly about the direction of **V**. We will make use of this technique in chapter 5 when we derive the proper form of the accelerations in terms of spherical coordinates.

EXERCISES

Exercise 2-1 [lines 10000–10290] *Velocity hodograph and acceleration plots.* This program draws position, velocity, and acceleration vectors similar to those shown in figure 2.2. It is not a very important program, but is perhaps amusing to look at. There are three panels. The left one is a plot of position, the middle of velocity, the right of acceleration. These quantities are computed for the case of a central force [line 10080] where K is an amplitude factor for the force and N is the power radius of the force [line 10060]. You can set $N = 1$ for simple harmonic motion, or $N = -2$ for Newtonian attraction toward the center. You can also set initial positions X, Y and velocities XD, YD as you want [line 10060]. The integration [lines 10090–10100] is short and simple but the arrow drawing displays require quite a lot of manipulation [lines 10110–10280], as well as the arrow drawing subroutine [lines 60000–60080].

What you see in this program is how the differences in the location of the R vector as time t goes through discrete steps DT are the velocity vectors (in the figure the directions of the velocity vectors in the left panel are not exactly the same as those shown magnified in the central panel because the step DT is finite and the velocity plot is drawn for the actual time $T + DT$ rather than from the difference shown in the left panel). This gives some idea of the errors for finite difference methods, and why with DT decreasing to the limit 0 you can make the calculus precise. It also shows a source of errors in finite difference calculations such as the forward differencing scheme used in the primitive numerical integration. The middle panel has the axes XD, YD, the right panel has axes XDD, YDD. There really isn't much that you can do with this display except play around with the initial conditions and force laws [line 10060]. You may decide that you would prefer to make a program up that works more like the figure described in the text. An example follows.

```
10000 'exercise 2-1;velocity & accelerat **
10010 SCREEN 1:     KEY OFF:CLS:COLOR 0,2
10020 DT=.01: FAC = 35:FFF=10: FFC = 20
10030 T = -DT
10040 LOCATE 2,5
10041 PRINT " x,y  velocity   acceleration"
10050 XO=60: XDO=150: XDDO=230
10051 YO=100:YDO=YO:YDDO=YDO
10060 X = 1: Y=1: XD=-.8
10061 YD=.5 :T=0: CA = 4 :K= 1:N=1
10070 R = SQR(X^2+Y^2)
10080 FX=-K*(X/R)*R^N   :FY =-K*(Y/R)*R^N
10090  XDD=FX: XD=XDD*DT+XD: X = XD*DT+X
10100  YDD=FY: YD=YDD*DT+YD: Y = YD*DT+Y
10110 COUNT = COUNT+1
10120 IF COUNT = 1 THEN GOTO 10150
10130 T = T+DT :CT=COUNT MOD 20
10140 IF CT=0 GOTO 10150 ELSE GOTO 10070
10150 AX=XO :AY=YO:LX=FAC*X
10151 LY = FAC*Y : AL=.4 : LH = 4 :CA = 1
10160 GOSUB 60000
10170 AX=XDO :AY=YDO:LX=FAC*XD
10171 LY = FAC*YD : AL=.4 : LH = 4 :CA=2
10180 GOSUB 60000
10190 AX=XDDO :AY=YDDO:LX=FAC*XDD
10191 LY = FAC*YDD: AL=.4 : LH = 4:CA=3
10200 GOSUB 60000
10210 IF COUNT = 1 THEN GOTO 10070
10220 AX=FAC*LASTX+XO:AY=FAC*LASTY+YO
10221 LX=FAC*(X-LASTX):LY=FAC*(Y-LASTY)
10222 AL=.4 : LH = 4 :CA = 2
10230 IF COUNT <50  THEN CA = 1 ELSE CA = 2
10240 GOSUB 60000
10250 AX=FAC*LASTXD+XDO:AY=FAC*LASTYD+YDO
10251 LX=FAC*(XD-LASTXD): LY = FAC*(YD-LASTYD)
```

```
10252 AL=.4 : LH = 4 :CA=3
10260 IF COUNT <30  THEN CA = 2 ELSE CA = 3
10270 GOSUB 60000
10280 LASTX=X: LASTY=Y:LASTXD=XD:LASTYD=YD
10290 GOTO 10070
```

Exercise 2-2 [lines 11000–11290] *A differently computed example.* This program is mostly the same as the previous one. You may simply want to renumber lines 10000–10290 to read 11000–11290, and then change only a few necessary lines. In line 11050 the value of *YO* is changed for graphical adjustment of the plot. There is no line 11060. The new part, which makes this program different from exercise 2.1, is in lines 11070–11090. You see that we have defined a particular function for *X* and *Y* as functions of *T*. We have also written down the analytically derived expressions for *XD, YD* and *XDD, YDD* in lines 11080–11090. The plotting is then simple: there is no integration. You can amuse yourself a little by trying various forms of functions in lines 11070–11090.

```
11000 'exercise 2-2; velocity & acceleration
11010 SCREEN 1:     KEY OFF:CLS:COLOR 0,2
11020 DT=.01: FAC = 35:FFF=10: FFC = 20
11030 T = -DT
11040 LOCATE 2,6
11041 PRINT " x,y  velocity   acceleration"
11050 XO=60: XDO=150: XDDO=230
11051 YO= 50:YDO=YO:YDDO=YDO
11070 X = T + .2*T^2 +.2*SIN(T): Y= .5*T^3
11080 XD= 1+.4*T+.2*COS(T):YD=1.5*T^2
11090 XDD=.4-.2*SIN(T)  :YDD=3*T
11110 COUNT = COUNT+1
11120 IF COUNT = 1 THEN GOTO 11150
11130 T = T+DT :CT=COUNT MOD 20
11140 IF CT=0 GOTO 11150 ELSE GOTO 11070
11150 AL=.4 : LH = 4 :CA = 1
11151 AX=XO :AY=YO:LX=FAC*X  : LY = FAC*Y
11160 GOSUB 60000
11170 AX=XDO :AY=YDO:LX=FAC*XD
11171 LY = FAC*YD : AL=.4 : LH = 4 :CA=2
11180 GOSUB 60000
11190 AX=XDDO :AY=YDDO:LX=FAC*XDD
11191 LY = FAC*YDD: AL=.4 : LH = 4:CA=3
11200 GOSUB 60000
11210 IF COUNT = 1 THEN GOTO 11070
11220 AX=FAC*LASTX+XO:AY=FAC*LASTY+YO
11221 LX=FAC*(X-LASTX) : LY = FAC*(Y-LASTY)
11222 AL=.4 : LH = 4 :CA = 2
11230 IF COUNT <50  THEN CA = 1 ELSE CA = 2
11240 GOSUB 60000
11250 AX=FAC*LASTXD+XDO:AY=FAC*LASTYD+YDO
11251 LX=FAC*(XD-LASTXD):LY=FAC*(YD-LASTYD)
11252 AL=.4 : LH = 4 :CA=3
11260 IF COUNT <30  THEN CA = 2 ELSE CA = 3
11270 GOSUB 60000
```

```
11280 LASTX=X: LASTY=Y:LASTXD=XD:LASTYD=YD
11290 GOTO 11070
```

Exercise 2-3 [lines 13000–13140] *Sling shot.* This program starts by showing a stone swinging around in a circular arc, pulled toward a center by a force [the arrow] due to an inextensible massless string. The calculation is done with a rather small *DT*, but displayed only every 40 integration steps [line 13090 has a MOD 40 in it, which of course can be changed if you like]. If you strike lower case "b" you break the string and the force turns off [line 13130]. It then flies in a straight line, but not radially away from the center. Its direction is tangential to its circular path at the time of breaking the string. Many people find this fact counterintuitive. Perhaps this is the mistake that Goliath made. There are lots of practical examples: the sparks that fly from a rotating grindstone are broken bits of steel and stone. They fly off tangentially, but not because the knife blade somehow deflects them in that direction.

```
13000 'Exercise 2-3; sling shot ***********
13005 CLS
13010 DT =.05: XD=3: YD= 0 :X=100:Y=100
13011 TH =3.14159/2 :F=(XD^2)/25 :SC=5
13020 XDD=-F*SIN(TH):YDD= F*COS(TH)
13030 XD=XDD*DT+XD:X=XD*DT+X
13031 YD=YDD*DT+YD:Y=YD*DT+Y
13040 COUNT=COUNT+1:CT = COUNT MOD 40
13050 IF XD=0  THEN TH = SGN(YD)*3.1415/2
13060 IF XD=0 THEN GOTO 13090
13070 TH = ATN(YD/XD)
13080 IF XD>0 THEN TH=TH ELSE TH = TH+3.1415
13085 IF CT=1 GOTO 13090 ELSE GOTO 13100
13090 CIRCLE(X,200-Y),2,2:PAINT(X,200-Y),2
13100 AX = X: AY= Y:LX=-YD*SC:LY=XD*SC
13101 LH = 2  :AL=.4:CA = 1
13110 IF CT=1 AND F>0 THEN GOSUB 60000 'arrow
13120  C$ = INKEY$
13130  IF C$="b" THEN F = 0
13140 GOTO 13020
```

Exercise 2-4 [lines 14000–14840] *Vector construction.* This program needs subroutine S-1 [60000 ff] to draw arrows. The program is entirely graphics so the main thing to discuss here is what it shows. It is an animated form of figure 2.5. The program is operated by successive pressings of key "c" {continue}. There is a page number at the top left that tells what has been drawn. Pressing "c" once brings page 1, showing a point in the x, y plane connected to the origin by the radius r of a polar coordinate system. Pressing "c" again brings up page 2. Here we see a vector depicted by an arrow. The tail of the arrow is located at the point in question; the rest of the arrow does not occupy x, y space. We can view the x, y components of this vector on page 3, and eliminate the vector itself on page 4. Now we draw two lines, one an extension of the radius r through the point in question, and the other perpendicular to this line, through the point. These define the directions of the sought for polar components of the vector. We can see them on page 5. On page 6 we place the x and y components back on the picture again, and drop perpendiculars from the arrow heads of each of these components (page 6) onto the line defining the direction of the angular component of the vector in the polar coordinates. Page 7 shows the contribution to the angular component from the y component. On page 8 we see the contribution to the angular component from the x component. On page 9 they are added together to produce the full value of the angular component of the vector. On page 10 we draw page 5 again, and this time, on page 11, we drop our perpendiculars onto the line defining the direction of the radial component of the vector in polar coordinates. On pages 12 and 13 we see successively the two parts that are contributed by the projections of the x and y components, and on page 15 the total summed up radial component. Page 16 shows both the polar components of the vector together, and on page 17 we can inspect the polar components and the original rectilinear components simultaneously. All that remains to complete the cycle of our story is to show again, on page 18, the original vector itself.

```
14000 'Exercise 2-4; vector construction ***
14005 SCREEN 1: COLOR 0,2: KEY OFF:CLS
14010 GOSUB 14240 :GOSUB 14210
14020 LINE(0,0)-(0,199),1
14030 CLS:LOCATE 1,1: PRINT NUMBER
14040 GOSUB 14240:GOSUB 14270:GOSUB 14210
14050 GOSUB 14300 :GOSUB 14210
14060 GOSUB 14330 :GOSUB 14210
14070 CLS:LOCATE 1,1:PRINT NUMBER:GOSUB 14240
14071 GOSUB 14270:GOSUB 14330:GOSUB 14210
14080 GOSUB 14380:GOSUB 14210
14090 GOSUB 14420: GOSUB 14210
14100 GOSUB 14530:GOSUB 14210
14110 GOSUB 14560:GOSUB 14210
14120 CLS:LOCATE 1,1:PRINT NUMBER:GOSUB 14240
14121 GOSUB 14270:GOSUB 14590:GOSUB 14210
14130 CLS:LOCATE 1,1:PRINT NUMBER:GOSUB 14240
14131 GOSUB 14270: GOSUB 14330
14132 GOSUB 14380: GOSUB 14210
14140 GOSUB 14620:GOSUB 14210
14150 GOSUB 14730:GOSUB 14210
14160 GOSUB 14760:GOSUB 14210
14170 GOSUB 14790:GOSUB 14210
14180 CLS:LOCATE 1,1:PRINT NUMBER:GOSUB 14240
14181 GOSUB 14270:GOSUB 14790
14182 GOSUB 14210
14190 CLS:LOCATE 1,1: PRINT NUMBER
14191 GOSUB 14240:GOSUB 14270:GOSUB 14790
14192 GOSUB 14590:GOSUB 14210
14200 GOSUB 14330:GOSUB 14210
14205 GOSUB 14300:GOSUB 14210
14210 A$=INKEY$
14220 IF A$ = "c" THEN GOTO 14900
14230 GOTO 14210
14240 LINE(0,0)-(0,199),1
```

```
14250 LINE(0,199)-(300,199),1
14260 RETURN
14270 LINE(0,199)-(100,150),2
14280 CIRCLE(100,150),1,2
14290 RETURN
14300 AX=100 :AY =  50: LX=50
14301 LY=80:AL=.4:CA=3:LH =4
14310 GOSUB 60000
14320 RETURN
14330 AX=100 :AY =  50: LX=50
14331 LY= 0:AL=.4:CA=3:LH =4
14340 GOSUB 60000
14350 AX=100 :AY =  50: LX= 0
14351 LY=80:AL=.4:CA=3:LH =4
14360 GOSUB 60000
14370 RETURN
14380 LINE(100,150)-(200,100),2
14390 LINE (100,150)-(50,50),2
14400 LINE(100,150)-(150,250),2
14410 RETURN
14420 X=150:Y=150
14430 FOR I = 1 TO 10
14440 X = X-4: Y=Y+2
14450 PSET(X,Y),2
14460 NEXT
14470 X=100:Y=70
14480 FOR I = 1 TO 8
14490 X = X-4: Y=Y+2
14500 PSET(X,Y),2
14510 NEXT
14520 RETURN
14530 AX=100:AY=50:LX=-33:LY=66:CA=1
14540 GOSUB 60000
14550 RETURN
14560 AX=100:AY=50:LX=10 :LY=-20:CA=1
```

```
14570 GOSUB 60000
14580 RETURN
14590 AX=100:AY=50:LX=-20:LY=43 :CA=1
14600 GOSUB 60000
14610 RETURN
14620 X=100:Y=70
14630 FOR I = 1 TO 16
14640 X = X+2:Y=Y+4
14650 PSET(X,Y),2
14660 NEXT
14670 X=150:Y=150
14680 FOR I = 1 TO 5
14690 X = X-2:Y=Y-4
14700 PSET (X,Y),2
14710 NEXT
14720 RETURN
14730 AX=100:AY= 50:LX=37:LY=17:CA=1
14740 GOSUB 60000
14750 RETURN
14760 AX=100:AY= 50:LX=42:LY=20:CA=1
14770 GOSUB 60000
14780 RETURN
14790 AX=100:AY= 50:LX=75:LY=37:CA=1
14800 GOSUB 60000
14810 RETURN
14820 GOSUB 14730:GOSUB 14210
14830 GOSUB 14300:GOSUB 14210
14840 STOP
14900 NUMBER = NUMBER + 1
14910 LOCATE 1,1
14920 PRINT NUMBER
14930 RETURN
```

CHAPTER III

Rotating coordinate frames

3.1 CORIOLIS FORCE

Suppose that there is a frame of polar coordinates r', ϕ' rotating around the origin of our old frame with constant angular velocity Ω. The primes on the coordinates denote the fact that these quantities are referred to the rotating reference frame. There are no real force components in the old system so $F_r = F_\phi = 0$ and the equations are simple. To transform to the new rotating coordinates we note that the new value or radius r' must be the same as r in the old frame, because the transformation is only one of rotation. On the other hand the new angle ϕ' is related to the old angle by the relation $\phi' = \phi - \Omega t$. Making the substitutions into the dynamical equations (2.14) in plane polar coordinates we obtain

$$\ddot{r} - (\dot{\phi}' + \Omega)^2 r = 0 \qquad (3.1)$$

$$r\ddot{\phi}' + 2\dot{r}(\dot{\phi}' + \Omega) = 0. \qquad (3.2)$$

Now we will rewrite these equations with the conventional terms that we recognize as accelerations in polar coordinates on the lefthand side, and put the remaining terms on the right hand side where forces usually are, and just to keep our minds straight we write them as "apparent forces" $F'_{r'}$, $F'_{\phi'}$.

$$\ddot{r} - r\dot{\phi}'^2 = F'_{r'} \equiv 2\Omega r\dot{\phi}' + \Omega^2 r \qquad (3.3)$$

$$r\ddot{\phi}' + 2\dot{r}\dot{\phi}' = F'_{\phi'} \equiv -2\Omega\dot{r}. \qquad (3.4)$$

Here we have a set of equations that contain terms on the right hand side that look, for all the world [the rotating world, that is], like forces, although an observer in absolute space would know that there aren't any. These forces have names, which helps to make them seem real. But they are just a result of trying to write the physics of a Newtonian system into a rotating [accelerated] reference frame.

The component of "force" $F'_{r'}$ has two terms. One is an additional centrifugal "force" $\Omega^2 r$ due to the system's rotation rate Ω in absolute space, acting only in the radial direction. You will note that it contains only the angular velocity of the rotating frame, Ω, not the full angular velocity of the particle $\dot{\phi} = \Omega + \dot{\phi}'$. The other term $2\Omega r\dot{\phi}'$ is the radial component of the Coriolis force. It is the product of a parameter 2Ω, which has come to be called the Coriolis parameter, with the tangential component of the relative velocity $r\dot{\phi}'$. The positive sign means that if the counterclockwise component of relative angular velocity $\dot{\phi}'$ of the particle and Ω are positive then the product $2\Omega r\dot{\phi}'$ represents an apparently outward radial force, to the right of the direction of the velocity. In the second equation there is only one term on the right-hand side. It is $-2\Omega\dot{r}$ which is the azimuthal component of the Coriolis force. It is the product of the Coriolis parameter and the radial component of the velocity $\dot{r}$.

3.2 MAGNITUDE OF THE CORIOLIS FORCE

For most work-a-day mechanical phenomena on the earth our physical formulation can ignore Coriolis "forces" and actually ignore the earth's rotation, imagining that it is a true inertial frame. Let us make some order of magnitude estimates of the size of these apparent forces, ignoring for the moment the important geometrical differences between a rotating horizontal plane and a rotating spherical earth.

The length of a sidereal day is 8.62×10^4 solar seconds so the angular velocity of the earth with respect to the stars is $\Omega = 7.29 \times 10^{-5} \sec^{-1}$. The distance of our habitation from the axis of rotation can vary from 0 to 6.38×10^6 meters, the equatorial radius of the earth. Let us suppose that we are driving an automobile sedately at 10 m $\sec^{-1}$ and that if we take our foot off the gas the car comes to rest by friction in 20 seconds, a deceleration of 0.5 m $\sec^{-2}$. Therefore each unit mass of the car and passenger must be exerting a force per unit mass of something like 0.5 m $\sec^{-2}$ on the road surface to keep moving at 10 m $\sec^{-1}$ against friction of the air, and rolling friction due to the compressibility of the tires. First let us compare this with the Coriolis force acting to deflect the car to the right, $2\Omega \dot{r} = 1.458 \times 10^{-4} \times 10 = 1.46 \times 10^{-3}$ m $\sec^{-2}$. This is a very small force compared to the force per unit mass exerted by the car on the road, the balancing forces of friction, etc. So we can afford to ignore it.

Now let us look at the additional centrifugal force

$$\Omega^2 r = (7.29)^2 \times 10^{-10} \times 6.38 \times 10^6$$

$$= 3.4 \times 10^{-2} \mathrm{msec}^{-2}.$$

This time we have found something not really negligible. It means that if the car were first at rest it would be moving away from the origin at 10 m sec^{-1} after about 5 minutes, in the absence of any friction. You can work out its terminal velocity assuming frictional force is proportional to some power of velocity. What this illustrates is that the apparent centrifugal force is substantial: why don't we notice it? It could be disastrous for cars parked without their brakes on.

On the real earth the equatorward component of centrifugal force would be balanced by the uphill grade of the 21 kilometer high bulge at the equator. An 0.3% grade is not really very small. Even freight locomotives have difficulty negotiating grades of 2% and have to be coupled up in pairs to negotiate mountain passes. The real earth is banked like an automobile raceway. It is a steep enough a banking to keep cars traveling around the world (at absolute speeds of up to 1000 miles an hour) from sliding to the equator. Although this bulge removes our need to consider the centrifugal force due to the earth's rotation, which depends upon relative position only, it has no direct effect on the Coriolis forces, which depend upon the relative velocities, and with which we continue to contend.

What we learn from this example is that when we think about physical problems in rotating frames of reference, and wish to restrict the number of virtual "forces" to Coriolis forces, we are going to have to balance the centrifugal force associated with the rotating frames of reference in some kind of way—analogous to the bulge of the earth. Otherwise we will not be able to have a population of particles at rest in the rotating frame, as we can on the surface of the earth.

3.3 CENTRIFUGAL AND CORIOLIS FORCES IN ROTATING RECTANGULAR COORDINATES

The rotating axis could have been introduced in rectilinear form rather than in polar form. Suppose that x, y denote the position of a particle in the absolute reference frame, and x', y' the position relative to a frame rotating about the origin with uniform angular velocity Ω, so that at any time t we have the relation

$$x = x' \cos\Omega t - y' \sin\Omega t,$$

and

$$y = x' \sin\Omega t + y' \cos\Omega t.$$

Now suppose that there are no forces in the system in absolute space, so that we know that $\ddot{x} = \ddot{y} = 0$. We dot the above expressions twice and gather the right hand side terms in the following sequence

$$\ddot{x} = 0 = (\ddot{x}' - 2\dot{y}'\Omega - x'\Omega^2)\cos\Omega t$$
$$- (\ddot{y}' + 2\dot{x}'\Omega - y'\Omega^2)\sin\Omega t,$$

and

$$\ddot{y} = 0 = (\ddot{x}' - 2\dot{y}'\Omega - x'\Omega^2)\sin\Omega t$$
$$+ (\ddot{y}' + 2\dot{x}'\Omega - y'\Omega^2)\cos\Omega t.$$

If we multiply the first by $\cos(\Omega t)$ and the second by $\sin(\Omega t)$ and add we obtain

$$\ddot{x}' - 2\dot{y}'\Omega - x'\Omega^2 = 0.$$

On the other hand multiplying the first by $-\sin(\Omega t)$ and the second by $\cos(\Omega t)$ and adding we obtain

$$\ddot{y}' + 2\dot{x}'\Omega - y'\Omega^2 = 0.$$

We now remember that we are observing phenomena from the moving system, so we prefer to keep only the $\ddot{x}'$, $\ddot{y}'$ terms on the lefthand side. Moving the remainder to the right, we have

$$\ddot{x}' = 2\Omega\dot{y}' + \Omega^2 x' + X'$$
$$\ddot{y}' = -2\Omega\dot{x}' + \Omega^2 y' + Y'.$$

We have added X' and Y' as real external forces, in case we want them later. Again, you can see that there are two additional terms on the right hand side besides the bona fide forces X', Y'. They are really a part of the acceleration, but in the spirit of the mental game we call them forces. The first terms on the right are the Coriolis forces in the relative rotating rectilinear system. They are products of the Coriolis parameter 2Ω and the relative velocity components $\dot{x}'$, $\dot{y}'$. The second terms are centrifugal forces due to the rotation of the coordinate system at angular speed Ω.

We will now proceed to construct a simple example of motion that, when viewed from a certain set of uniformly rotating axes, has only the Coriolis forces on the right hand side, and eliminates the centrifugal term $\Omega^2 r$ exactly. In this example it will be possible to have a population of particles at rest in the rotating system. We'll call them experts. So take a breath—and we're going back across the gangplank again to absolute space, where we'll construct our system.

Suppose that the x, y plane is horizontal in absolute space, that z is vertically upward, and that gravity acts downward, in the z-negative direction. We fabricate a smooth frictionless plate whose top surface lies within the x, y plane. Particles of unit mass are held against the top surface of the plate by gravity, which in turn is exactly balanced by the upward reaction of the plate acting upon the particle. There is, thus, always a balance of forces in the z-direction and the particle remains forever in the x, y plane. It is perfectly free to move in the x, y plane, because the surface of the plate is smooth, flat, frictionless. The plate can move in the x, y direction, or around a vertical axis, without having any influence whatever on the x, y dynamics of the particle. The only effect of the plate and gravity is to restrict the degrees of freedom of motion of the particle to two instead of three dimensions, so that z is always equal to zero.

3.4 EXPERTS, NOVICES AND HOOKE SPRINGS

Now we attach the particle of unit mass to the origin by means of a massless spring whose tension is equal to $-k^2 r^n$ where k^2 is the so-called spring constant, and n can be any power whatever. If we choose $n = 1$ then the spring is an approximation to a real spring, whose tension increases with extension r, while if we choose $n = -2$ the spring models an additional gravitational attraction toward the origin.

The equations of motion in polar form in absolute space for the general central force $F_r = -k^2 r^n$ are

$$\ddot{r} - \dot{\phi}^2 r = -k^2 r^n,$$

and

$$r\ddot{\phi} + 2\dot{r}\dot{\phi} = 0.$$

The radial acceleration depends upon $-k^2 r^n$; and the angular momentum does not change. Circular orbits around the origin are always possible provided the initial conditions are chosen expertly: $\dot{r}_i = 0$: $\dot{\phi}_i^2 = -k^2 r_i^{n-1}$. Then $\ddot{r} = \dot{r} = \ddot{\phi} = 0$ forever. We shall denote the values of r and ϕ that refer to these expertly chosen initial conditions by a subscript e, thus r_e, ϕ_e. When a particle is started around the axis as an *expert* it always goes in a circle. When a particle is started as a *novice* it generally follows some other path—we will omit the subscript for an experiment conducted by a novice. You can imagine that an expert is a person (particle) with slippery shoes who grabs a spring and jumps on to the rotating system with just the correct angular velocity so that he slides around the axis to appear at rest relative to the rotating reference frame; but a novice jumps on with an absolute angular velocity or radial velocity that is not in equilibrium with the rotating reference frame so that he appears, relative to that frame, to be moving. We will frequently refer to *experts* and *novices* in this context in this book. It is a somewhat unusual nomenclature, but is clear, unambiguous, anthropomorphic, and rather intuitive in meaning. It gives the abstract problem the concrete feeling of a skating rink.

Now inspection of the relation $\dot{\phi}_e^2 = k^2 r^{n-1}$ shows us that the only choice of n that allows all experts to circle around the axis with the same angular velocity independent of the radius r is the choice $n = 1$, which corresponds to the so-called Hooke spring. With $n = 1$, a cloud of experts executes its rotation in solid rotation. The shape of the cloud, the relative position of each particle in the cloud of experts, when viewed from the point of view of an observer rotating with angular velocity $\dot{\phi}_e$ does not change. In fact the entire cloud appears to be at rest in the rotating reference frame, unless the observer happens to glance upward to the stars—which he will see in apparent rotation about the axis.

The artifice of tying every unit mass of the system to the center with an identical Hooke spring allows us to eliminate that extra centrifugal force $\Omega^2 r$ due to the rotation of the rotating reference frame in equation 3.3. Now, there will be no apparent force acting upon the particles at rest in the rotating system, and they will remain at rest. They will actually rotate in solid rotation about the axis in absolute space. They will constitute a "platform," made up of "expert" particles. We might express the concept in other language: the experts are a configuration of particles in rotational equilibrium. Particles that are not at rest in the rotating reference frame (our "novices" who jumped on to it with springs but not with proper choice of initial velocity) find themselves acted upon by the Coriolis forces.

3.5 TRAJECTORIES IN THE ABSOLUTE INERTIAL REFERENCE FRAME

We now must ask ourselves about fact and fiction when we observe particles started on our plate as novices, so that r and ϕ change with time in some more complicated way. And because many of us are a little uncomfortable with the polar form of the equations, we will revert temporarily to the equations in rectilinear form in the absolute frame. Because $F_x = F \cos \phi$, $F_y = F \sin \phi$ and $\cos \phi = x/r$, $\sin \phi = y/r$, and $\Omega = \dot{\phi}_e = k$ [here we are setting the rotation rate equal to the square root of the spring constant as before], these equations are simply [with choice $n = 1$]:

$$\ddot{x} = -\Omega^2 x, \qquad \ddot{y} = -\Omega^2 y. \qquad (3.5a)$$

They do look more comfortable than the polar coordinate form, and they are simply the equations in the absolute frame for simple harmonic motion, with the familiar general solutions

$$x = A_1 \cos \Omega t + B_1 \sin \Omega t$$

$$y = A_2 \cos \Omega t + B_2 \sin \Omega t$$

where the constants A_1, A_2, B_1, and B_2 are determined by the initial conditions x_i, y_i, $\dot{x}_i$, $\dot{y}_i$ at $t = 0$. Thus:

$$x = x_i \cos \Omega t + (\dot{x}_i/\Omega) \sin \Omega t \qquad (3.5b)$$
$$y = y_i \cos \Omega t + (\dot{y}_i/\Omega) \sin \Omega t.$$

The trajectories are all ellipses in the x, y plane with center at the origin. They all have the same period $2\pi/\Omega$.

The expert point moves in a circle because, for example, if it starts at $x = r_e$, $y = 0$ with initial velocities $\dot{x}_i = 0$, and $\dot{y}_i = \Omega r_e$ we have simply the equation for a circle

$$x_e = r_e \cos \Omega t, \qquad y_e = r_e \sin \Omega t.$$

We can construct a whole cloud of experts that rotate around the origin in circles of various radii in solid rotation. Novices move in elliptical paths.

3.6 A LINKAGE ANALOG

The elliptical path of the novice particle and the path of an expert moving in a circle in absolute space can be reproduced kinematically by a simple linkage generating epicycles.* Consider a rod of length a, pivoted freely at the origin, with angle Ωt measured counterclockwise

---

*See *Dynamics,* by Horace Lamb (Cambridge: Cambridge University Press, 1961), pp 62 ff.

from the x axis. Its free end has coordinates $x = a \cos \Omega t$, $y = a \sin \Omega t$. Pivoted to the free end is another rod of length b, with angle $2\Omega t$ measured clockwise from the first rod, as shown in figure 3.1. The two rods therefore always have equal but opposite slope. The coordinates of the point P at the free end of rod b are therefore

$$x = a\cos\Omega t - b\cos\Omega t \equiv (a - b)\cos\Omega t$$

$$y = a\sin\Omega t + b\sin\Omega t \equiv (a + b)\sin\Omega t.$$

These equations define an elliptical orbit in absolute space, and the rod a may be thought of as the relative x' axis. They mimic the motion described in equations (3.5b) for the special case $\dot{x}_i = y_i = 0$. Obviously simple rearrangements of the linkage can mimic other cases.

As the angle Ωt varies from 0 to 2π the point P therefore describes an ellipse with minor axis $(a - b)$ in the x direction, and a major axis $(a + b)$ in the y-direction and center at the origin, and period $2\pi/\Omega$. In the x, y space the point P also describes a circle about the moving center C with the same period, but in the opposite direction from that with which point C rotates with time. We now mark the inner rod at the point where P touches it when $t = 0$ with a point P_e. Obviously it describes a circle around the center with radius $a - b$ in the same period $2\pi/\Omega$.

$$x_e = (a - b)\cos\Omega t, \qquad y_e = (a - b)\sin\Omega t.$$

If we now place ourselves in the position of an observer riding around this latter circle with the expert point P_e, facing away from the origin, we will be in the rotating frame of apparently stationary experts, but the point P will now appear to us to rotate around a circle clockwise two times for every actual rotation of the expert P_e. The

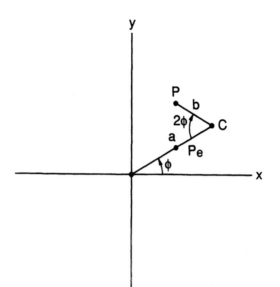

Figure 3.1. The linkage that demonstrates how a point P seems to make a complete circle about C when viewed by an observer (the expert) riding on a rod a at P_e, while the angle ϕ has swept out only half a complete revolution.

apparent period of the novice P will be only one-half the period of the reference frame with respect to absolute space. This factor of two comes up all the time when we encounter the Coriolis force, as we are now about to do, as we transfer ourselves to the rotating frame.

3.7 TRAJECTORY IN ROTATING FRAME

What we have demonstrated above, in a special case for simplicity, is an epicyclic trajectory that has the same functional form as a particular solution in equations (3.4) and hence can be used to describe it easily, especially when transforming to the moving axes.

At this point we return to the dynamical problem of the unit mass tied to the origin with a Hooke spring with spring constant, k^2 and a rotating reference frame rotating with angular velocity $\Omega = k$.

The polar forms of the equations of particle motion in the absolute frame are

$$\ddot{r} - \dot{\phi}^2 r = -k^2 r = -\Omega^2 r,$$

and   (3.6a)

$$r\ddot{\phi} + 2\dot{r}\dot{\phi} = 0.$$

As we have seen already we can show that the experts rotate as a solid if $\dot{\phi} = \Omega$, independently of r and ϕ. Now we will move to the rotating frame of the experts, r', ϕ', where $\phi' = \phi - \Omega t$. Making this substitution we obtain the equation in the rotating frame

$$\ddot{r}' - (\dot{\phi}' + \Omega)^2 r' = -\Omega^2 r'$$ (3.6b)

$$r'\ddot{\phi}' + 2\dot{r}'(\dot{\phi}' + \Omega) = 0.$$

The terms $-\Omega^2 r'$ cancel out from the equations. In other words, the tension of the Hooke spring balances the centrifugal force due to the Ω rotation of the reference system. Now we write the relative equations in conventional form with accelerations on the left-hand side of the equation and the remainder of the terms on the right-hand side.

$$\ddot{r}' - \dot{\phi}'^2 r' = 2\Omega r' \dot{\phi}'$$
$$r'\ddot{\phi}' + 2\dot{r}\dot{\phi}' = -2\Omega \dot{r}' \tag{3.7}$$

The term $r'\dot{\phi}'$ is the relative azimuthal component of velocity $v_{\phi'}$ and the term $\dot{r}'$ is the relative radial velocity $v_{r'}$, so we can rewrite the equations as

$$\ddot{r}' - \dot{\phi}'^2 r' = 2\Omega v_{\phi'}$$
$$r'\ddot{\phi}' + 2\dot{r}\dot{\phi}' = -2\Omega v_{r'}. \tag{3.8}$$

We can, of course, immediately transform these polar equations in the moving reference frame to the rectilinear equations in the moving reference frame

$$\ddot{x}' = 2\Omega \dot{y}', \quad \text{and} \quad \ddot{y}' = -2\Omega \dot{x}'. \tag{3.9}$$

Again we see the Coriolis forces, this time as product of the same Coriolis parameter and the corresponding rectilinear velocity components. At first glance there seems to be no trace of the Hooke spring constant, but of course the spring is present implicitly through its determination of the quantity $\Omega = k$. It is important to note that these equations [unlike equations (3.5a)] are linked together because each contains x' and y'.

Even though these Coriolis forces are not real forces, they provide a handy way of doing computations in the relative frame. For a novice that starts at the same position as the expert the solution of these equations is

$$x' = x'_i + (\dot{y}'_i/2\Omega)(1 - \cos 2\Omega t) + (\dot{x}'_i/2\Omega)\sin 2\Omega t$$

$$y' = y'_i - (\dot{x}'_i/2\Omega)(1 - \cos 2\Omega t) + (\dot{y}'_i/2\Omega)\sin 2\Omega t$$

where everything is written in terms of the moving reference frame. The initial velocity of the expert in the moving frame is, of course, zero. For the novice these equations describe a circular trajectory, with center at

$$x'_e = x'_i + (\dot{y}'_i/2\Omega), \qquad y'_e = y'_i - (\dot{x}'_i/2\Omega).$$

The radius of the circle is $[(\dot{y}'_i)^2 + (\dot{x}'_i)^2]^{1/2}/2\Omega$. The period is π/Ω. The sense of rotation in the rotating frame is in the same sense (clockwise) as the apparent motion of the stars, but twice as fast.

Suppose that x points eastward and y northward and $\Omega > 0$. Then, according to equation (3.9), the Coriolis force due to a northward relative velocity, $\dot{y}' > 0$, produces an eastward relative acceleration, $\ddot{x}' > 0$ and the Coriolis force due to a westward relative velocity, $\dot{x}' < 0$, produces a northward relative acceleration, $\ddot{y}' > 0$. The Coriolis force acts to the right of the momentary direction of relative velocity and does no work.

3.8 ANOTHER APPROACH USING COMPLEX NOTATION

The material we have just presented is so fundamental that it is worth exploring again using yet another approach. Let us retain the subscript i to stand for the initial value of a quantity, but let the letter $i = \sqrt{-1}$ and use complex notation to explore the geometry once more. Denote the position of the particle in the plane by the complex number $x + iy$. We can also write

$$e^{i\Omega t} = \cos\Omega t + i\sin\Omega t,$$

from which it follows that

$$\cos\Omega t = \frac{1}{2}[e^{i\Omega t} + e^{-i\Omega t}]; \qquad \sin\Omega t = -\frac{i}{2}[e^{i\Omega t} - e^{-i\Omega t}].$$

Then equations (3.5b) can be written together as

$$x + iy = (x_i + iy_i)\cos\Omega t + \frac{\dot{x}_i + i\dot{y}_i}{\Omega}\sin\Omega t$$

$$= \frac{1}{2}(x_i + iy_i)(e^{i\Omega t} + e^{-i\Omega t}) - \frac{i(\dot{x}_i + i\dot{y}_i)}{2\Omega}(e^{i\Omega t} - e^{i\Omega t})$$

$$= \frac{1}{2}\left[\left(x_i + \frac{\dot{y}_i}{\Omega}\right) + i\left(y_i - \frac{\dot{x}_i}{\Omega}\right)\right]e^{i\Omega t}$$

$$+ \frac{1}{2}\left[\left(x_i - \frac{\dot{y}_i}{\Omega}\right) + i\left(y_i + \frac{\dot{x}_i}{\Omega}\right)\right]e^{-i\Omega t}.$$

Since a complex position of the form

$$x + iy = Ae^{i\Omega t} = |A|e^{i\alpha}e^{i\Omega t}$$

is to be interpreted as a vector of length $|A|$ oriented at angle α to the x axis at time $t = 0$, and rotating counterclockwise around the origin at frequency Ω, the first bracketed term describes the basic circular trajectory of the point C in figure 3.1. The condition that this be the total motion in the absolute inertial frame means that the coefficient of $e^{-i\Omega t}$ should be zero, which means that

$$\dot{y}_i = \Omega x_i, \qquad \dot{x}_i = -\Omega y_i.$$

But these are simply the conditions satisfied by the "expert" particles, $x_e + iy_e = (x_i + iy_i)e^{i\Omega t}$ being the expert trajectory. This is not exactly the same expert particle we had before since we started at an arbitrary point $x_i + iy_i$ on the trajectory rather than the endpoint of the semi-minor axis of the ellipse, P_e.

Subtracting this from the actual particle path gives a trajectory relative to the rotating reference system, which is

$$x + iy - (x_e + iy_e) = \frac{1}{2}[e^{-i\Omega t} - e^{i\Omega t}]\left[\left(x_i - \frac{\dot{y}_i}{\Omega}\right) + i\left(y_i + \frac{\dot{x}_i}{\Omega}\right)\right].$$

This relative trajectory viewed from the expert position by an observer always facing away from the origin means the factor $e^{i\Omega t}$ in the absolute inertial frame must be reduced to $1 = e^{i0}$ in the moving frame. So we must multiply the above result by $e^{-i\Omega t}$ to obtain

$$x' + iy' = \frac{1}{2}[e^{-i2\Omega t} - 1]\left[\left(x_i - \frac{\dot{y}_i}{\Omega}\right) + i\left(y_i + \frac{\dot{x}_i}{\Omega}\right)\right],$$

or simply,

$$x' + iy' = C[e^{-i2\Omega t} - 1],$$

where C is a complex constant. Differentiating twice with respect to time yields

$$(\ddot{x}' + i\ddot{y}') = -i2\Omega(\dot{x}' + i\dot{y}'),$$

and writing the real and imaginary parts of this gives

$$\ddot{x}' = 2\Omega \dot{y}', \qquad \ddot{y}' = -2\Omega \dot{x}'.$$

These are the rectilinear equations in the moving reference frame, derived by putting the solutions in the moving frame and seeing what equations they satisfy, rather than by transforming the differential equations directly. The result, of course, is the same, but the process is analogous to getting from the wharf to the ship by crossing two different gangplanks. You end up in the same place either way.

3.9 THE USAGE OF THE WORDS "BALANCE" AND "EQUILIBRIUM"

Perhaps this is a good place to say something about our usage of the words "balance" and "equilibrium" in order to avoid a common source of confusion.

Balance generally refers to a balance of forces. In the case of a cloud of experts, each tied to the axis by a Hooke spring, and each revolving around the axis at different radii but with the same constant angular velocity, it is abundantly clear that the forces acting on each particle are not in balance. In fact there is only one force acting on each: that of the tension of the stretched spring. There are no other forces to afford a balance. The force (per unit mass of particle) is proportional to the radius, and is precisely that required to accelerate each particle toward the axis so that each particle can revolve about the axis with the same angular velocity Ω.

Thus the cloud of experts rotates as a solid body: the distance between each expert and its neighbors remains constant. So we can think of the cloud as being in a sort of equilibrium—an equilibrium configuration to be explicit. But this is not the same as a balance of forces. Therefore when we use the words balance and equilibrium it is worth asking ourselves to be a little more explicit: a "balance" or "equilibrium" of what?

The imbalance of forces in an equilibrium configuration of uniform rotation with Hooke springs is particularly simple to understand because there is only one force acting on each particle. In later chapters, where equilibrium configurations of uniform rotation of an assembly of experts involve an imbalance of more than one force, it is a little more complicated. We will then have to consider combinations of gravitation and the reaction of smooth slippery surface on a particle, or finally, when we prepare to think of fluid continua, the imbalance will involve gravitation and pressure gradients.

Novice particles are not part of the equilibrium configuration: they move relative to the experts.

If we transform our point of view to a reference frame that is fixed relative to the rotating equilibrium configuration of experts, all the experts will appear to be at rest. If we insist on thinking that we want our dynamics to be similar to that in an inertial resting reference frame (the Newtonian one), then we can write down an apparent balance of forces by introducing the virtual centrifugal force to balance the spring's tension on each expert particle. This is clearly an artifice introduced for convenience of calculation in the rotating reference frame.

When we want to write down equations for the motion of novices relative to the experts in the rotating reference frame we must introduce some additional virtual (unreal) forces: the Coriolis forces. Again this is something of a semantic trick that often proves useful for producing a notationally economical form of equations of motion.

PROBLEMS

Problem 3.0. The sea-ice at the North Pole is frozen into a perfectly smooth frictionless surface. How fast must a puck be driven to execute an inertial circle 200 meters in diameter?

Problem 3.1. It is often useful to interpret phenomena as small perturbations from an equilibrium state. Suppose that in absolute space, on our slippery plane, with Hooke springs, a particle is rotating at constant equilibrium radius r_e with angular velocity Ω. Now suppose that it is slightly perturbed so that it departs from the equilibrium state by a small amount indicated by the prefix delta: $r = r_e + \delta r$, $\phi = \phi_e + \delta\phi$. Write the polar coordinate form of the equations in terms of this sum, subtract the equilibrium part, and neglect terms with products of the delta terms. Interpret the resulting perturbation equations physically, giving meaning to the perturbation Coriolis terms that appear.

Problem 3.2. A particle initially at rest on the rotating Hooke plane is attached to another point fixed on the plane by a Hooke spring that is very stiff. What is the qualitative nature of the orbit of the particle in the rotating reference frame?

Problem 3.3. Two experts [numbered 1 and 2] on the slippery plane [each holding his own Hooke spring] are rotating about the axis at angular velocity Ω. They are at radii r_1, r_2 and angles ϕ_1, ϕ_2. At the time $t = 0$, $\phi_1 = 0$ and the expert 1 drops a puck of small mass to the platform. We consider two cases.

Case 1. The puck is not attached to a Hooke spring, so

in absolute space no force acts on it in the horizontal. It therefore slides in a straight trajectory. If the angle of expert 2 at time $t = 0$ is ϕ_2, then at what time will he be able to pick up the puck as it slides under his feet? How does the feasibility of such a pickup depend upon the two radii r_1 and r_2, the rotational angular velocity of the reference frame Ω and the angle ϕ_{20}? The discussion of this problem is best formulated in absolute space because there is no spring attached to the puck.

Case 2. Now suppose that the puck has an appropriate spring attached. When dropped by expert 1 it just stays at his feet. What is the minimum velocity relative to the moving reference frame that he must throw it in order for it to reach the expert 2? If he misses, it will always return to him, so that he can try again. Are there other choices of magnitude and direction of relative velocity that he can throw it at, in order to reach expert 2? This case is best formulated in the uniformly rotating coordinate system, because the puck [with spring attached] is subject only to Coriolis forces.

EXERCISES

Exercise 3-1 [lines 15000–15260] *Particle on flat platform with center-connected springs.* This exercise is for the purpose of illustrating the simple harmonic motion that occurs on a flat plane with a centrally directed force due to Hooke-type springs. Time step *DT,* the radius of the expert *RE,* the spring constant *K,* and a small quantity *EPS* that is used for setting the amplitude of the novice's error in judging the correct angular velocity (*PED*) to jump onto the platform with are all given default values in line 15020. The program first asks you to type "c" [line 15030]. This line has a loop that is generating random numbers *ZZZZ* so that you'll get a different start with the novice each time. Once "c" is calculated you go to 15050 where the screen is cleared and some labelling is done on the screen. The angular velocity for an expert at *RE* is *PED*, calculated in 15090, and the novice's angular velocity *PD* is set randomly about *PED* in 15100. The integration is done [lines 15120–15150] using the polar coordinate form of the equations for the simple harmonic case discussed in chapter 3. The remainder of the program is graphics. If you don't want to see trajectories, but only the instantaneous points, remove the *REM* ['] symbols after the line numbers in lines 15160, 15170, and 15200. This causes an overwrite with color 0 [black] on the last plotted positions *RL*, *PL*, *REL*, *PEL*, where the suffix *L* means last. As set up, the *PE* trips line 15230 when the platform has made one full turn, and goes to 15260 where the phases are reset, and then goes back to 15050, clearing the screen. A completely new independent run, with a different value of ratio of novice to expert initial angular velocities, is printed on the screen by line 15110.

You should watch this program run through several times, to see the different ellipses generated in the absolute space (shown on the left panel) and by the two particles' positions in the reference frame that rotates with the equilibrium velocity of the expert (shown in the right panel). The circles on the right always pass through the expert position, always complete a rotation in half the time it takes the ellipses on the left to do so. When *YD/ YED* > 1 the ellipse is larger than the expert's circle, and the circle in the rotating reference frame lies outside the radius of the expert. When *YD/YED* lies between 0 and 1, the ellipse in absolute space lies inside the expert circle, and the inertial circle in the rotating reference frame lies inside the expert's radius. You will find it interesting to try *YD/YED* < 0. Think about what you expect, and then try it out. Explain what you see.

```
15000 ' Exercise 3-1,rotating platform *******
15001 ' with springs to center
15010 SCREEN 1:COLOR 0,2: KEY OFF: CLS
15020 DT=.01:PI=3.14159:RE=20:K=1:EPS=3
15030 ZZZZ=RND: A$=INKEY$
15031 LOCATE 10,13: PRINT "Type c"
15040 IF A$ = "c" THEN GOTO 15050
15041 GOTO 15020
15050 CLS
15060 LOCATE 20,5
15061 PRINT " absolute        relative"
15070 CIRCLE(80,100),50,1
15080 CIRCLE(220,100),50,1
15090 PED= SQR(K)
15100 RD=0:R=RE:PD=EPS*(RND-1/2)+PED
15105 QA=PD/PED
15110 LOCATE 4,8
15111 PRINT USING "initial pd/ped= ##.###";QA
15120 RDD = R*PD^2 - K*R: PDD=-2*RD*PD/R
15130 RD=RDD*DT+RD: PD=PDD*DT+PD
15140 R = RD*DT+R: P=PD*DT+P
15150 PE=PED*DT+PE
15180 PSET( 80+R *COS(P),100-R *SIN(P)),3
15190 PSET(220+R*COS(P-PE),100-R*SIN(P-PE)),3
15210 PSET(220+RE,100),2
15220 PSET(80+RE*COS(PE),100-RE*SIN(PE)),2
15230 IF PE>2*PI THEN GOTO  15260
15240 RL=R:PL=P:REL=RE:PEL=PE
15250 GOTO 15120
15260 P = 0: PE= 0 : GOTO 15050
```

Exercise 3-2 [lines 16000–16250] *Epicycle display.* This program is a display in motion of figure 3.1 in the text. It shows three panels: on the left is the trajectory of an expert and novice particle in simple harmonic motion on the *X, Y* plane—absolute reference frame. On the right is the trajectory of the same two particles on the rotating reference frame *MX, MY.* In the center two linked rods rotate as the text describes (at the top, in absolute space, and at the bottom in the moving coordinate system), and the two particles are shown by heavy dots. The excess of tangential velocity of the novice over the expert is controlled by the parameter *B* in line 16020. Most of the program is simple graphics and doesn't require much discussion here. What you want to do with this program is to watch the two large dots (particles) on the epicycle and try to imagine that you are riding on the inner rod at the point where the expert is. You are turned with the inner rod at the rate of the moving axes, and so should be able to see the novice dot make the inertial circle. This program ought to help you make that intuitive link between the motions as depicted on the left panel to those depicted on the right panel.

```
16000 'Exercise 3-2;Epicycle diagram *******
16010 SCREEN 1: COLOR 0,2: KEY OFF: CLS
16020 FAC = 30: B = 2: DT = .1
16030 LOCATE 2,3
16031 PRINT "absolute    epicycle   relative"
16040 XE= COS(T): YE= SIN(T)
16050 X = COS(T): Y=B*SIN(T)
16060 XC=((B+1)/2)*COS(T):YC=((B+1)/2)*SIN(T)
16070 PSET(40+FAC*X,100-FAC*Y),2
16080 PSET(40+FAC*XE,100-FAC*YE),3
16090 MX = X *COS( T)+Y *SIN(T)
16100 MY = -X*SIN(T) +Y *COS(T)
16110 PSET(200+FAC*MX,100-FAC*MY),2
16120 T = T+DT
16130 GOSUB 16150
16140 GOTO 16040
16150 'plots
16155 IF T > 6.28 THEN CLS
16156 IF T > 6.28 THEN T=0
16160 LINE(80,20)-(220,200),0,BF
16162 FOR J = 0 TO 1
16163 IF J = 1 THEN XE = 1 ELSE XE - XE
16164 IF J = 1 THEN L = -2.2 ELSE L = 0
16165 IF J = 1 THEN YE=L ELSE YE = YE
16166 IF J = 1 THEN X = MX ELSE X=X
16167 IF J = 1 THEN Y = L+MY ELSE Y=Y
16168 IF J = 1 THEN XC=(B+1)/2 ELSE XC=XC
16169 IF J = 1 THEN YC=L ELSE YC=YC
16170 XX=150+FAC*X:YY=100-FAC*Y
16175 YL=100-FAC*L: XXC=150+FAC*XC
16180 YYC=100-FAC*YC
16187 LINE( XX,YY)-(XXC,YYC),1
16188 LINE( XX,YY)-(XXC,YYC),1
16189 LINE( 150,YL)-(XXC,YYC),1
16190 LINE( 150,YL)-(XX,YY ),1
```

```
16200 CIRCLE(XXC,YYC),FAC*(B-1)/2,2
16210 FOR I = 0 TO 3
16220 CIRCLE(150+FAC*XE,100-FAC*YE),I/2,3
16230 CIRCLE(150+FAC*X ,100-FAC*Y ),I/2,3
16240 NEXT :NEXT
16250 RETURN
```

Exercise 3-3 [lines 17000−17260] *Astronaut jumps toward center of rocket.* The cylindrical body of a spaceship is rotating about its axis with angular velocity Ω. This provides a "centrifugal force" that feels like gravity to an astronaut walking around on the inside of the cylinder. At time $t = 0$ he jumps vertically (toward the axis). How will his trajectory look in absolute space, and in the frame of the cylindrical ship—that is, as viewed by other astronauts who haven't jumped off the inside platform?

The radius of the ship's hull A, rate of rotation OMEGA, time increment DT (note that it has to be rather small to avoid integration errors near the axis, and therefore the display is often slow) and a graphical magnification factor FAC are assignable [lines 17030−17040]. Drawing the two views of the section through the ship occupies lines 17050−17100. The basic integration starts in a loop at 17110 where RDD is computed from a form of the radial acceleration equation in which the angular variable PD has been replaced by substitution of constant angular momentum MU [line 17020]. Integration procedes through 17150, and is then converted to coordinates for display and plotting to 17190. The intercepts [17200,17210] stop the display when the astronaut lands on the inner wall of the ship again. Lines 17220−17260 draw rather heavily the location of the jumping off place. We see that this marked part of the hull rotates in the lefthand panel in absolute space, but stands still in the frame relative to the rotating ship on the right panel.

As should be expected, the path of the astronaut is a straight line in absolute space, but a curve in the moving frame of the ship.

```
17000 'Exercise 3-3: astronaut jumps ********
17010 SCREEN 1: COLOR 0,2: KEY OFF: CLS
17020 A = 1: OMEGA = 3: MU= OMEGA*A^2
17030 R = A: RD = -20: DT = .0001:FAC = 50
17040 PD=OMEGA     : AP = A
17050 FOR I = 1 TO 4 STEP .5
17060 CIRCLE( 80,100),50+I,1
17070 CIRCLE( 80,100),  I,1
17080 CIRCLE(220,100),50+I,1
17090 CIRCLE(220,100),  I,1
17100 NEXT
17110 RDD=MU^2/R^3
17120 RD=RDD*DT+RD: R= RD*DT+R
17130 PDD=-2*RD*PD/R:PD=PDD*DT+PD:P=PD*DT+P
17140 X=R*COS(P):Y=R*SIN(P)
17150 T=T+DT
17160 MX=R*COS(P-OMEGA*T):MY=R*SIN(P-OMEGA*T
17170 XE=(AP )*COS(OMEGA*T)
17171 YE=(AP)*SIN(OMEGA*T)
17180 PSET( 80+FAC* X,100-FAC* Y),3
17190 PSET(220+FAC*MX,100-FAC*MY),3
17200 IF R>A THEN BEEP
17210 IF R>A THEN END
17220 CIRCLE(80+FAC*XE,100-FAC*YE),2,2
17230 PAINT (80+FAC*XE,100-FAC*YE),2
17240 CIRCLE(220+FAC*A ,100      ),2,2
17250 PAINT (220+FAC*A ,100      ),2
17260 GOTO 17110
```

Exercise 3-4 [lines 18000–18870] *Billiards on the Hooke spring Coriolis plane.* This program is a rather complicated one, involving conditional statements for collisions among three balls and the side cushions. Dr. James Luyten wrote this version of the program, and kindly gave his permission for us to include it in this book. Because the program is long and complicated we will not describe its instructions line by line. Instead, we simply explain how to use it.

After loading it, you will be requested to choose a value for the Coriolis parameter. We advise you to begin with 0, so that you will first have an ordinary game of billiards to contend with. Then type one of the cursor control keys. The ↑ key rotates the cue counterclockwise about the cue ball, the ↓ key rotates the cue clockwise. The ← key diminishes the strength of the stroke of the cue, the → key increases it. Then type "g" to begin the game.

There are some interesting features to the game when a modest COR is specified, say COR = .02. The balls tend to be deflected to the right, so that it is not entirely trivial to produce collisions without cushion shots. Balls often get in a bouncing mode along the cushions, in which they progress counterclockwise around the rim of the table [reminiscent of Kelvin waves to the cognoscenti]. Of course it is possible to have a situation in which both the balls have inertial circles that are so small and isolated from cushions and each other that there never can be any collisions—but you might ask can one of these be set up ab initio by striking the cue ball?

```
18000 '3-4  Billiards with Coriolis force ****
18002 'By courtesy of James Luyten
18005 ' the x,y velocities are xid,yid
18010 ' the amplitude of the velocities are mi
18015 ' ti,angles of velocities in x,y coords
18020 ' ph, angle of tangent when circles touch
18025 ' the rotated velocities are ui,vi
18030 ' bi,angles of velocity,in rotated coords
18035 ' the ni are dummies for mi
18040 ' COR is the Coriolis parameter
18045 ' Use cursor keys to initialise
18050 ' Type <g> to begin the game
18055 TINY =1D-20:SKE1=0:SKE2=0:SKE3=0
18060 DIM CUE%(500),BALL%(500),OTHER%(500)
18065 DT0=2 :R=10 : PI = 3.14159265#
18070 SCREEN 1: COLOR 0,2: KEY OFF: CLS
18071 WINDOW SCREEN (-50,-50)-( 900, 500)
18075 KEY(11) ON:KEY(12) ON
18076 KEY(13) ON :KEY(14) ON
18080 CIRCLE (100,100),R,3:PAINT (100,100),3
18081 GET (80,80)-(120,120),CUE%
18085 CIRCLE (100,100),R,2:PAINT (100,100),2
18086 GET (80,80)-(120,120),OTHER%
18090 CIRCLE (100,100),R,1:PAINT (100,100),1
18091 GET (80,80)-(120,120),BALL%
18095 CLS
18100 LOCATE 1,1
18101 INPUT "Coriolis parameter(0 to .1)";COR
18105 LINE(0,0)-(300,399),1,B
18110 Z = 180/PI   'radians to degrees
18115 X1= 50:Y1=100:X2=150:Y2=100
18116 X3=250:Y3=225
18120 PUT (X1-20,379-Y1),CUE%
18125 PUT (X2-20,379-Y2),BALL%
18130 PUT (X3-20,379-Y3),OTHER%
18135 M1= 5:M2=0:M3=0:T1 =-10/Z
18136 T2= 0/Z:T3=0/Z:S=M1/600 's normalises
```

```
18140 X10=X1:X20=X2:Y10=Y1
18141 Y20=Y2 :Y30=Y3:X30=X3
18145 GOSUB 18755 'get initial conditions
18150 M10=M1:T10=T1:GOSUB 18715
18155 DT=DT0
18160 'integration of dynamical equations
18165 Y1DD=-COR*X1D :Y1D=Y1D+Y1DD*DT
18170 Y2DD=-COR*X2D :Y2D=Y2D+Y2DD*DT
18175 Y3DD=-COR*X3D :Y3D=Y3D+Y3DD*DT
18180 X1DD=COR*Y1D :X1D=X1D+X1DD*DT
18185 X2DD=COR*Y2D :X2D=X2D+X2DD*DT
18190 X3DD=COR*Y3D :X3D=X3D+X3DD*DT
18195 X1=X1D*DT+X1: Y1=Y1D*DT+Y1
18200 X2=X2D*DT+X2: Y2=Y2D*DT+Y2
18205 X3=X3D*DT+X3: Y3=Y3D*DT+Y3
18210 M12= X1D^2+Y1D^2:M1=SQR(M12)
18211 X=X1D:Y=Y1D:GOSUB 18625:T1 =THETA
18215 M22 = X2D^2+Y2D^2:M2=SQR(M22)
18216 X=X2D:Y=Y2D:GOSUB 18625:T2 =THETA
18220 M32 = X3D^2+Y3D^2:M3=SQR(M32)
18221 X=X3D:Y=Y3D:GOSUB 18625:T3 =THETA
18225 GOTO 18255'interactions befor plotting
18230 T=T+DT
18235 GOSUB 18620
18240 GOSUB 18660            'plot
18245 X10=X1:X20=X2:X30=X3
18246 Y10=Y1:Y20=Y2:Y30=Y3
18250 GOTO 18160
18255 IF X1>=300-R THEN GOTO 18335 'cushions
18260 IF X2>=300-R THEN GOTO 18335 'cushions
18265 IF X3>=300-R THEN GOTO 18335 'cushions
18270 IF X1<= R THEN GOTO 18335
18275 IF X2<= R THEN GOTO 18335
18280 IF X3<= R THEN GOTO 18335
18285 IF Y1<=R THEN GOTO 18335
18290 IF Y2<=R THEN GOTO 18335
18295 IF Y3<=R THEN GOTO 18335
```

```
18300 IF Y2>=399-R THEN GOTO 18335
18305 IF Y1>=399-R THEN GOTO 18335
18310 IF Y3>=399-R THEN GOTO 18335
18315 DI3=((X2-X1)^2+(Y2-Y1)^2)/ (2*R)^2
18316 IF DI3<1 THEN GOTO 18335 ' collision
18320 DI2=((X3-X1)^2+(Y3-Y1)^2)/ (2*R)^2
18321 IF DI2<1 THEN GOTO 18335 ' collision
18325 DI1=((X2-X3)^2+(Y2-Y3)^2)/ (2*R)^2
18326 IF DI1<1 THEN GOTO 18335 ' collision
18330 GOTO 18230
18335 AAA=RECALC
18340 IF AAA=1 THEN GOTO 18365 ELSE RECALC=1
18341 DT=DT0/3
18345 X1=X10:Y1=Y10'reset xi,yi to recalcul
18350 X2=X20:Y2=Y20
18355 X3=X30:Y3=Y30
18360 GOTO 18160
18365 IF X1>=300-R THEN X1D=-X1D    'cushions
18370 IF X2>=300-R THEN X2D=-X2D
18375 IF X3>=300-R THEN X3D=-X3D
18380 IF X1<= R THEN X1D=-X1D
18385 IF X2<= R THEN X2D=-X2D
18390 IF X3<= R THEN X3D=-X3D
18395 IF Y1<=R THEN Y1D=-Y1D
18400 IF Y2<=R THEN Y2D=-Y2D
18405 IF Y3<=R THEN Y3D=-Y3D
18410 IF Y2>=399-R THEN Y2D=-Y2D
18415 IF Y1>=399-R THEN Y1D=-Y1D
18420 IF Y3>=399-R THEN Y3D=-Y3D
18425 DI3=((X2-X1)^2+(Y2-Y1)^2)/ (2*R)^2
18426 IF DI3<1 THEN GOSUB 18470 ' collision
18430 DI2=((X3-X1)^2+(Y3-Y1)^2)/ (2*R)^2
18431 IF DI2<1 THEN GOSUB 18520 ' collision
18435 DI1=((X2-X3)^2+(Y2-Y3)^2)/ (2*R)^2
18436 IF DI1<1 THEN GOSUB 18570 ' collision
18440 DT=DT0:RECALC=0
18445 X=X1D:Y=Y1D:GOSUB 18625:T1 =THETA
```

```
18450 X=X2D:Y=Y2D:GOSUB 18625:T2 =THETA
18455 X=X3D:Y=Y3D:GOSUB 18625:T3 =THETA
18460 GOTO 18160
18465 RETURN
18470 NUM = NUM+1 ' collision
18475 BEEP
18480 X=X2-X1:Y=Y2-Y1:GOSUB 18625
18481 ALPHA=THETA:SA=SIN(ALPHA):CA=COS(ALPHA)
18485 U1N=M1*COS(ALPHA-T1) 'velocities in
18486 U1T= M1*SIN(ALPHA-T1) 'normal coords
18490 U2N=M2*COS(ALPHA-T2)
18491 U2T= M2*SIN(ALPHA-T2)
18495 V1N=U2N:V2N=U1N:V1T=U1T:V2T=U2T
18500 ' transformation back into x,y coords
18505 X1D=V1N*CA+V1T*SA:Y1D=V1N*SA-V1T*CA
18510 X2D=V2N*CA+V2T*SA:Y2D=V2N*SA-V2T*CA
18515 RETURN
18520 NUM = NUM+1 ' collision
18525 BEEP
18530 X=X3-X1:Y=Y3-Y1:GOSUB 18625
18531 ALPHA=THETA:SA=SIN(ALPHA):CA=COS(ALPHA)
18535 U1N=M1*COS(ALPHA-T1)
18536 U1T= M1*SIN(ALPHA-T1)
18540 U3N=M3*COS(ALPHA-T3)
18541 U3T= M3*SIN(ALPHA-T3)
18545 V1N=U3N:V3N=U1N:V1T=U1T:V3T=U3T
18550 ' transformation back into x,y coords
18555 X1D=V1N*CA+V1T*SA:Y1D=V1N*SA-V1T*CA
18560 X3D=V3N*CA+V3T*SA:Y3D=V3N*SA-V3T*CA
18565 RETURN
18570 NUM = NUM+1 ' collision
18575 BEEP
18580 X=X2-X3:Y=Y2-Y3:GOSUB 18625
18581 ALPHA=THETA:SA=SIN(ALPHA):CA=COS(ALPHA)
18585 U3N=M3*COS(ALPHA-T3)
18586 U3T= M3*SIN(ALPHA-T3)
18590 U2N=M2*COS(ALPHA-T2)
```

```
18591 U2T= M2*SIN(ALPHA-T2)
18595 V3N=U2N:V2N=U3N:V3T=U3T:V2T=U2T
18600 ' transformation back into x,y coords
18605 X3D=V3N*CA+V3T*SA:Y3D=V3N*SA-V3T*CA
18610 X2D=V2N*CA+V2T*SA:Y2D=V2N*SA-V2T*CA
18615 RETURN
18620 RETURN
18625 'atn2(y,x)
18630 IF X=0 THEN THETA=SGN(Y)*PI/2 : RETURN
18635 IF ABS(X)<TINY  THEN X=SGN(X)*TINY
18640 THETA=ATN(Y/X)
18645 IF X<0 THEN THETA=THETA+PI
18650 IF X>0 AND Y<0 THEN THETA =THETA+2*PI
18655 RETURN
18660 ' plot balls
18665   PUT (X10-20,379-Y10),CUE%
18670 IF IP=1 THEN PSET (X1,400-Y1),3
18675   PUT (X1-20,379-Y1),CUE%
18680   PUT (X20-20,379-Y20),BALL%
18685 IF IP=1 THEN PSET (X2,400-Y2),1
18690   PUT (X2-20,379-Y2),BALL%
18695   PUT (X30-20,379-Y30),OTHER%
18700 IF IP=1 THEN PSET (X3,400-Y3),2
18705   PUT (X3-20,379-Y3),OTHER%
18710 RETURN
18715 'cls&change history
18720 IP=(IP+1) MOD 2:CLS
18721 LINE(0,0)-(300,399),1,B:
18725 PSET (X1,400-Y1),6:PSET (X2,400-Y2),3
18726 PSET (X3,400-Y3),5
18730 PUT (X1-20,379-Y1),CUE%
18735 PUT (X2-20,379-Y2),BALL%
18740 PUT (X3-20,379-Y3),OTHER%
18745 LOCATE 1,1
18746 PRINT USING "initial: m1=###.##";M10
18747 LOCATE 1,20 :AA=T10*Z
18748 PRINT USING "t1=###.# COR=.##";AA,COR;
```

```
18750 RETURN
18755 'get initial conditions
18760 ON KEY(11) GOSUB 18790
18765 ON KEY(12) GOSUB 18795
18770 ON KEY(13) GOSUB 18800
18775 ON KEY(14) GOSUB 18805
18780 IF INKEY$="g" THEN GOTO 18835
18785 GOTO 18760
18790 T1=T1+1/Z:GOSUB 18810:RETURN
18795 M1=M1-1 :GOSUB 18810:RETURN
18800 M1=M1+1:GOSUB 18810:RETURN
18805 T1=T1-1/Z:GOSUB 18810:RETURN
18810 X1D=M1*COS(T1): Y1D=M1*SIN(T1)
18811 X2D=M2*COS(T2):Y2D=M2*SIN(T2)
18815 AAA=X1+30*X1DO:YYY=399-Y1-Y1DO*30
18816 LINE(X1,399-Y1)-(AAA,YYY),0'erase old
18820 AAA=X1+30*X1D:YYY=399-Y1-Y1D*30
18821 LINE(X1,399-Y1)-(AAA,YYY),7
18825 X1DO=X1D:Y1DO=Y1D:X2DO=X2D:Y2DO=Y2D
18830 LOCATE 1,1
18831 PRINT USING "initial: m1=###.##";M1
18832 LOCATE 1,19:AA=T1*Z
18833 PRINT USING "t1=####.# COR=.##";AA,COR;
18835 RETURN
18840 X1D=X1D+TINY*RND
18845 Y1D=Y1D+TINY*RND
18850 X2D=X2D+TINY*RND
18855 Y2D=Y2D+TINY*RND
18860 X3D=X3D+TINY*RND
18865 Y3D=Y3D+TINY*RND
18870 RETURN
```

SOME PHYSICAL INTERPRETATION OF WHAT WE HAVE OBSERVED IN EXERCISE 3-1

The two most important exercises in this book are 3-1 and 7-1. Both illustrate motions in rotating systems with platforms that are arranged so that they balance the centrifugal forces due to rotation of the system. Thus it is possible for a universe of particles to move in solid rotation about the axis of rotation, so that an observer living on the rotating frame would be unconscious of the existence of the Hooke springs (in 3-1) or the equatorial bulge of the earth (in 7-1). When he observes particles in motion relative to the rotating frame, the Coriolis forces rather mysteriously appear. We have seen how they arise formally. Let us now attempt a verbal, physical explanation of the phenomena we have observed in the two panels on the left and right when we ran exercise 3-1. The program has been set up to introduce random fluctuations in the angular velocity *PD* of the novice particle. It will be more convenient, if we want to perform controlled experiments, to be able to set values for the novice's radial and angular velocity at will. So the easiest way to do this is to insert for line 15100, the following

15110   INPUT "*RD,PD*"; *RD,PD*

This supersedes previous assignments, and the values now assigned are displayed on the screen.

The spring constant and rotation rate are set up so that the values for the expert or equilibrium particle are taken as $RE = 20$ initial radius, $RED = 0$ initial radial velocity, $PE = 0$ initial angle, $PED = 1$ initial angular velocity. This particle moves in equilibrium around a circle in absolute space, and appears to be at rest in rotating space. We now consider the novice, or disturbed particle. We set it at the same initial position, $R = 20$, $P = 0$, but we assign various values of RD and PD to it. The following are some experiments, and physical interpretations of them.

(1) Let us begin by disturbing the novice particle by giving it a larger than equilibrium radial velocity, but the same choice of angular velocity $PD = PED$, for example say $RD = 10$, $PD = 1$. This means that in the absolute space the particle begins by moving radially away from the expert particle. It has the same angular momentum as the expert, however, so now the novice, being further from the axis of rotation, must have a smaller angular velocity, and begins to fall behind the expert in azimuth. The smaller angular velocity, however, means that it cannot quite balance the inward radial force of the Hooke springs (larger at large radius) and so it begins to accelerate toward the axis again. Its radial velocity reverses, it approaches its starting radius again, but because of its loss in azimuth during its excursion to large radius it passes the radius of the expert some distance in azimuth behind it. During the next half of its disturbed path relative to the expert, the novice now overshoots its initial radius so that it is now closer to the axis than the expert. By conservation of angular momentum its angular velocity must now be greater than that of the expert, so now the novice, during the time it is inside the radius of the expert, gains in azimuth and catches up with the expert. The excess angular velocity, however, increases the centrifugal force in the radial dynamical equation so the particle now reverses radial velocity again and commences to move back outward toward the expert. It meets the expert after half a rotation of the reference frame in absolute space. Although it is not obvious from this description, the path of the disturbed novice particle on the moving platform is a perfect circle. However, it is obvious that the sense of mo-

tion of the novice about the circle relative to the rotating population of experts is clockwise, that the circle passes through the expert position and lies entirely in that portion of *MR, MP* space that has a azimuth less than that of *MPE*. This latter position is because, with the initial conditions we assumed in this example, the azimuth of the novice begins to fall behind that of the expert at the very beginning of the experiment, and is always less than that of the expert except when it passes over it.

(2) In this experiment we input $RD = -10$, $PD = 1$. This time the novice particle begins by moving inward. As it does so, conserving angular momentum, it gains azimuth over the expert. Now it finds its centrifugal force too great for the springs, so its radial velocity reverses and it starts to move back toward the radius of the expert. This time its azimuth is greater than that of the expert, so it passes in front of it. During the next half of its excursion at a radius greater than that of the expert, it falls back in azimuth again, and eventually is pulled back by the springs to pass over the position of the expert. As observed in the moving platform the circle is described in the same sense as before, in the same time, but this time lies everywhere at an azimuth greater than that of the expert, except for the point where expert and novice coincide. You may want to repeat these two experiments with $RD = 20$ and -20 instead of 10, -10 just to see what happens.

(3) Now we try another experiment, this time disturbing the angular velocity of the novice particle instead of the radial velocity. This means that the novice will always have an angular momentum that is different from that of the expert. Does this alarm you? Are you thinking that now we have come to a case where the disturbed particle simply must wander away from the expert slowly in time. After all, the novice cannot shed its excess angular momentum.

So now input $RD = 0$, $PD = 2$. Remember $PED = 1$, so you have given the novice a really big shot of angular momentum. It immediately begins to shoot out ahead of the expert in azimuth. But it finds that its centrifugal force is not fully balanced by those springs, and it begins to acquire an outward radial velocity. As it does so, its angular velocity diminishes (by conservation of angular momentum) and its azimuth begins to fall back to that of the expert, although of course it is at a larger radius. Eventually its angular velocity is less than that of the expert, its centrifugal force becomes weaker than that of the springs (which of course is larger at larger radius), and the springs begin to pull the novice back in toward the center again. It passes through the expert again. This time you will notice that the orbit of the novice has been entirely outside the radius of the orbit of the expert. It is still a circle as seen from the moving platform, but the circle is never inside the radius of the expert. Now how about that conundrum about the novice drifting away from the expert? You see that the novice spends nearly all its time at radii greater than that of the expert, so even though its angular momentum is greater than that of the expert, its average angular velocity is not.

(4) The fourth experiment is almost redundant. You may think that because it is just the opposite of experiment (3) you know what it is and what it will show. Let us input $RD = 0$, $PD = 0$. Now the novice has no angular momentum at all! So how can it possibly keep up with the expert? All the novice can do is to swing back and forth across the origin in response to the Hooke springs—a simple harmonic oscillator. But watch what happens. Surprisingly it meets the expert again after half a rotation, and again the orbit, as seen from the moving frame, is a circle that intersects both the original expert position and the axis. This example takes a little getting used to—

especially that business about the angular momentum. Maybe you should try some nearby inputs, for example INPUT $RD = 0$, $PD = .2$, or $RD = 0$, $PD = -.2$.

(5) Finally we will do one more experiment that will perhaps surprise you. Input $RD = 0$, $PD = -1$. Now we have set the angular momentum and velocity of the novice to be exactly opposite to that of the expert. The centrifugal forces depend upon squares of angular velocity, so they don't notice any difference. The springs don't either. So the novice goes around the axis in a perfect circle in absolute space, just as the expert does, but in opposite directions. Now look at the system from the point of view of the observer on the rotating platform. He sees the particle doing one of its famous inertial circles, this time around the axis.

(6) Now let us try one final experiment. We input $RD = 0$, $PD = -2$. We get a pretty large inertial circle this time, but look at what the observer sees as an apparent deflection in the moving reference frame. The expert is at rest, of course. The novice moves rapidly in the negative direction. At the initial instant its radial velocity is of course zero, but shortly thereafter it is positive. As an observer on the rotating platform, would you interpret this as an example in which the Coriolis force was apparently toward the left? or is this explicable some other way? Think over the previous examples, and ask the same question.

CHAPTER IV

The paraboloidal dish

4.1 THE PARABOLOID AS A PLATFORM

One's sense of simplicity is offended by the presence of all those Hooke springs needed to support the elements of the platform and the objects sliding around on it that we have just discussed. Hooke springs on a flat plane have the great advantage of simple analysis, of course, and the motions are two-dimensional. It may occur to readers that much the same kind of centripetal acceleration could be provided to a sliding object by erecting a stationary platform in the form of a large, fixed, curved dish, concave upward, with a very slippery surface. If we introduce the third dimension, z, to represent the vertical with respect to our constant downward gravity g in the absolute reference frame, then a dish of paraboloidal form can be written with the expression $z = cr^2$. This is what we get by rotating a parabola around the vertical axis at the origin of the x, y plane. There are two forces that act upon any object sliding on the dish: gravity acting downward, and a reactive force acting normal to the surface of the dish (figure 4.1). If we write the amplitude of this reactive force as F, and its radial and vertical components as F_r and F_z then, taking the angle of slope as α, which is simply $\alpha = \tan^{-1}(dz/dr)$, our equations of motion for the object

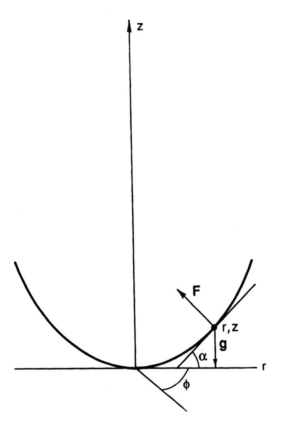

Figure 4.1. Cylindrical coordinates r, ϕ, z that are used in the discussion of the parabolic (paraboloidal) dish. The gravity g is downward. The reaction F of the dish's surface on the particle is normal to the surface of the dish, because the dish is slippery and cannot support any tangential force. The angle ϕ lies in the plane $z = 0$.

sliding on the dish, in absolute inertial space of three dimensions, are

$$\ddot{r} - \dot{\phi}^2 r = -F\sin\alpha,$$
$$r^2\ddot{\phi} + 2r\dot{r}\dot{\phi} = 0, \qquad (4.1)$$

and

$$\ddot{z} = -g + F\cos\alpha.$$

In these expressions the reactive force F is still to be determined, the fourth equation being $z = cr^2$. From this we have $\dot{z} = 2cr\dot{r}$ and $\ddot{z} = 2c(\dot{r})^2 + 2cr\ddot{r}$. Now we can substitute this expression for $\ddot{z}$ into the vertical equation of motion and obtain

$$2c(\dot{r}^2 + r\ddot{r}) = -g + F\cos\alpha. \qquad (4.2)$$

We can now eliminate the unknown reaction F from the system between this equation and the one for the radial acceleration to obtain the following two equations for calculating $\ddot{r}$ and $\ddot{\phi}$ of the object sliding on the dish in absolute space at rest to the stars.

$$\ddot{r} = \frac{-2cr(g + 2c\dot{r}^2) + \dot{\phi}^2 r}{1 + (2cr)^2},$$

and $\qquad\qquad\qquad\qquad\qquad\qquad\qquad$ (4.3)

$$r\ddot{\phi} = -2\dot{r}\dot{\phi}.$$

In the case where c is very small (a shallow dish) the terms with square of c get very small and this system approaches the one of chapter 3, yielding a solution for simple harmonic motion, and the previous expressions for Coriolis forces, etc.

Sometimes motorcycle race courses are designed in the form of a paraboloidal bowl. Motorcycles at different radii will have to travel at speeds proportional to the radius to avoid slipping sideways: thus those near the rim need to travel faster to stay "banked up" near the rim. But for points that move at constant radius the angular velocity is the same, so experts riding in circles around the axis of the parabolic dish do in fact all rotate in solid rotation. The concept of a reference platform is preserved. However, radial oscillations do not have periods independent of amplitude, and the motion in general is not simple harmonic.

4.2 SMALL AMPLITUDE MOTIONS IN THE ROTATING FRAME

We now will investigate motions on a paraboloidal surface that are small in amplitude about a radius r_0 and about a uniform angular velocity Ω. In concrete terms we are considering a novice who is very nearly as skillful as an expert who is moving steadily around the axis nearby.

Let us write $r = r_0 + r'$ and $\phi = \Omega t + \phi'$ in the equations (4.3), and linearize the resulting system. The primed parts of these expressions are the small deviations of the novice's coordinates from those of the expert in absolute inertial space. By the word linearize we mean that we intend to ignore all quadratic and higher degree terms in the primed quantities. The second equation of (4.3) becomes

$$(r_0 + r')\ddot{\phi}' = -2\dot{r}'(\Omega + \dot{\phi}'),$$

which linearizes to

$$r_0\ddot{\phi}' = -2\Omega\dot{r}'. \tag{4.4}$$

The first equation of (4.3) becomes

$$[1 + 4c^2(r_0 + r')^2]\ddot{r}'$$
$$= -2c(r_0 + r')[g + 2c\dot{r}'^2] + (r_0 + r')(\Omega + \dot{\phi}')^2$$

There are two large terms in this expression that have no primed quantity in them. They represent the equilibrium of the expert, and of course balance

$$-2cr_0g + r_0\Omega^2 = 0,$$

or

$$\Omega^2 = 2cg.$$

This defines the equilibrium Ω in terms of g and the shape of the paraboloid c. The first order terms (those with first degree primed quantities) are

$$(1 + 4c^2r_0^2)\ddot{r}' = 2\Omega r_0\dot{\phi}', \qquad (4.5)$$

where some r' terms on the right hand side dropped out because $\Omega^2 = 2cg$. We may integrate (4.4) once to obtain $r_0\dot{\phi}' = -2\Omega r'$ and substitute this result into equation (4.5). Then we have

$$(1 + 4c^2r_0^2)\ddot{r}' = -4\Omega^2r', \qquad (4.6)$$

which is the equation for simple harmonic motion at frequency $\omega = 2\Omega/\sqrt{1 + 4c^2r_0^2}$. At $r = r_0$ let the angle α in figure 4.1 be called α_0. Then $\tan\alpha_0 = dz/dr|_{r = r_0} = 2cr_0$, so the frequency of the small amplitude oscillation is

$$\omega = 2\Omega\cos\alpha_0.$$

Inasmuch as α_0 is the angle between the normal to the paraboloid and the z axis (of rotation) it is analogous to a geographic colatitude on an ellipsoidal earth (chapter 9). So the frequency of the small amplitude simple harmonic oscillation of a novice about the position of a neighboring expert on the paraboloid is just $\omega = 2\Omega \sin \vartheta$ where ϑ is equivalent to the geographic latitude. We will find that this same law of frequency for small amplitude oscillations of a particle about equilibrium also holds on a ellipsoidal earth.

Let us look a little more closely at the nature of motion. The perturbation of radius is

$$r' = \gamma \sin \omega t,$$

where γ is an arbitrary small amplitude. Accordingly we can write

$$r_0 \dot{\phi}' = -2\Omega\gamma \sin \omega t,$$

so

$$r_0 \phi' = \frac{\gamma}{\sin \vartheta} \cos \omega t.$$

Thus we see that on the plane $z = 0$ the trajectories are ellipses. On the other hand we can construct a plane tangent to the paraboloid at the expert [point $r' = \phi' = 0$]. In this plane we let $x' = r_0 \phi'$ measure the distance along the plane in the increasing ϕ' direction and $y' = r'/\sin \vartheta$ measure distance along the plane in the increasing r' direction. In these new x', y' coordinates the trajectories are circles:

$$x'^2 + y'^2 = \gamma^2/\sin^2 \vartheta.$$

We will also encounter this result when we consider the small motions of a particle on a rotating spheroid. On the other hand it is important to compare this solution to that for the Hooke spring plane, for which all novice orbits viewed from the rotating frame are circles—even when the amplitude is large—and for which the frequency is always constant regardless of the position on the plane.

Radial motion on the paraboloidal dish introduces vertical accelerations, which in turn affect the reactive force F. The component of this force then departs from simple dependence upon r alone, so that in general the paraboloidal system is not simple-harmonic except near to the axis of rotation where the slope of the dish and the induced vertical accelerations are small, or in the case of small amplitude perturbations from expert equilibrium.

4.3 FIRST INTEGRALS

It is instructive at this point to consider the general question of first integrals of the equations for motion on a surface of revolution. Let the surface of revolution be given by an equation of the general form $z = f(r)$. Then the slope of the surface at any point is $\tan \alpha = dz/dr = f'(r)$ where $f'(r)$ is the derivative of $f(r)$ with respect to r. The condition that the particle moves on the surface means that $\dot{z} = f'(r)\dot{r} = \tan \alpha \cdot \dot{r}$. Then equations (4.1) can be combined in the following way. Multiply the first by $\dot{r}$, the second by $\dot{\phi}$, and the third by $\dot{z}$ and add to obtain

$$\dot{r}(\ddot{r} - \dot{\phi}^2 r + F\sin\alpha) + \dot{\phi}(r^2\ddot{\phi} + 2r\dot{r}\dot{\phi}) \tag{4.7}$$
$$+ \dot{z}(\ddot{z} + g - F\cos\alpha)$$

$$= \dot{r}\ddot{r} + r^2\dot{\phi}\ddot{\phi} + r\dot{r}\dot{\phi}^2 + \dot{z}\ddot{z} + + g\dot{z} = 0.$$

The terms containing the normal constraining (reactive) force F vanish identically because $\dot{z} = \tan \alpha \cdot \dot{r}$ as long as the particle is on the surface. Of course, if the particle leaves the surface we no longer have $\dot{z} = \tan \alpha \cdot \dot{r}$, but if the particle leaves the surface there is no reactive force F either so therefore equation (4.7) is still true. Equation (4.7) can be written in the form

$$\frac{d}{dt}\left(\frac{\dot{r}^2}{2} + \frac{r^2\dot{\phi}^2}{2} + \frac{\dot{z}^2}{2} + gz\right) = 0,$$

which means

$$\dot{r}^2 + r^2\dot{\phi}^2 + \dot{z}^2 + 2gz = \frac{2E}{m}, \tag{4.8}$$

where the constant E is called the total energy of the particle. This is used as a starting point in the next chapter. Equation (4.8) follows from equation (4.7) regardless of the dependence of the individual variables r, ϕ, and z on time and thus represents a constraint on the motion of the particle. A relation of this kind is called an integral (or in particular, a first integral) of the equations of motion. There is another first integral of the equation (4.1). The second equation may be written as

$$\frac{d}{dt}(r^2\dot{\phi}) = 0,$$

and integrated to give

$$r^2\dot{\phi} = \frac{A}{m} = \mu, \tag{4.9}$$

where the constant A is called the angular momentum of

the particle, m is the mass of the particle, and μ the angular momentum per unit mass.

The two constants E and A (or μ) are set by the initial conditions at the start of the motion. Given $\dot{r}_i$, $\dot{\phi}_i$ and r_i [and therefore $\dot{z}_i = f'(r_i)\dot{r}_i$], E and A can be evaluated and remain constant thereafter.

Now, if the original equations are numerically integrated by forward time-stepping, as is done in the exercises, the equations (4.8) and (4.9) can be used as a check on the accuracy of the integration. But there is an alternative use, which we will explore.

PROBLEMS

Problem 4.0. Two small billiard balls are at rest relative to a shallow rotating paraboloidal billiard table. Their coordinates r, ϕ are different. How will you choose an impulse on the cue ball that will ensure that it will strike the other? How will inclusion of vertical acceleration affect this choice?

Problem 4.1. Show that if a particle is projected onto the inner surface at the rim of any concave dish, with a tangential velocity, that it will never pass through the lowest central point of the dish.

Problem 4.2. A pendulum bob is suspended by a massless rod. The system is spun around a vertical axis with increasing angular velocity. At what critical rate of spin can the bob begin to move away from the vertical, and at higher spins, what angle will it subtend from the vertical?

EXERCISES

Exercise 4-1 [lines 20000–20650] *Paraboloidal dish.* This program displays the movement of an expert particle, a novice particle and a third particle that is computed ignoring vertical acceleration, and therefore may be called the "approximate" particle, on a paraboloidal dish. Gravity acts downward. The display offers two options: with *OPT* = 0 the display on the left is in the *R, P* plane, that is, the absolute coordinate system but on the right the same particles are shown in the *MR, MP* system relative to the rotating reference frame; with *OPT* = 1 the display is shown in an oblique side view so that the curvature of the dish is visible. The dish can be tipped from the vertical different amounts by changing *INC* [inclination] in line 20030. The default *INC* = .4 radians. With *INC* = 0 the *OPT* = 1 view is directly from the side; with *INC* = *PI*/2, the view is a plane one.

Parameters of the problem are given default values in lines 20100 and 20140. The integration of the equations (4.4) are performed in lines 20170–20270. Transformation to *X, XE, XA* [the *A* is for approximate solution] and *Y, YE, YA* are in lines 20280–20340. The remainder of the program is graphics.

When you run the program with default values the *PD* is given a substantially larger angular velocity than the expert [an amount *EPS* as shown in line 20140—this can be altered at will]. The approximate particle is given the same, *PAD* = *PD*. The novice and approximate particle both run ahead of the expert around the dish, and mount up the steep side. In this initial stage the novice runs ahead of the approximate particle. Then they are slowed down in angular velocity—by conservation of angular momentum, if you like to think of it this way—and at their larger radii begin to fall behind the expert, so that as they run down hill again they are behind the expert. The approximate particle continues around in a circle—the typical inertial circle of chapter 3 on the flat Hooke spring plane. However, the novice falls even farther behind, and executes a cycloid-type trajectory, gradually receding toward lesser P (toward the west by analogy to the earth). The effect of the vertical acceleration terms in the dynamics of the paraboloidal dish are such as to retard novices as they proceed around the dish. This may perhaps seem reasonable when one recalls that while all experts go around a paraboloid with the same period, particles that go in the R, Z plane with $P =$ constant all have longer periods than the experts, and this period increases with the amplitude of the radial excursion. It seems plausible that oscillations which oscillate in modes that involve changes of both R and P should exhibit periods longer than those of experts.

As we remember from chapter 3, a disturbed novice particle always hovered about the expert that was at the starting point, and passed over it twice per revolution of the rotating platform in absolute space. On the other hand, in the parabolic dish the disturbed particle falls behind the expert in absolute azimuth, and in the rotating reference frame appears to drift away from the expert in the clockwise direction—in the meantime, of course, executing an epicycle-like trajectory. We can demonstrate this fact and make a simple physical interpretation in the special case where we compare the trajectories of the expert particle with a novice that is started at rest in absolute space. To do this we must set $PD = 0$ in line 20150. The rest of the line should remain the same. Now when we run this program the expert will be seen to circle the origin in the left hand panel, while it appears to be at rest in the righthand panel.

The novice that is computed by neglecting vertical acceleration RA, PA etc., will move straight across the origin in absolute space, much as a linear pendulum, and it will meet the expert at the far left end of its swing. In the relative panel the approximate novice executes an inertial circle passing through the expert and the origin—repeating the circuit every half rotation of the axes. On the other hand the novice particle (R, P, etc.) that does not neglect vertical acceleration is not able to keep up with the approximate novice because as it leaves the initial position it begins to fall, and this vertical acceleration reduces the vertical component of the reaction of the surface of the dish against the particle. As a result, the geometry being unaffected, the component of reaction toward the axis is also reduced and the particle is not accelerated as quickly toward the axis as its approximate companion. So it begins to fall behind. Once it passes the axis the particle begins to accelerate upward, and the particle's horizontal velocity is reduced to zero at the radius of the expert. But the delay in making this swing brings the exact novice to the end of its swing too late to meet the expert. As a consequence, in the righthand panel, where the trajectories are drawn relative to the moving axes, the trajectory of the exact novice is a series of loops gradually drifting away from the expert in a clockwise direction.

```
20000 ' Exercise 4-1; paraboloidal dish ******
20010 INPUT "plan (0),or oblique view(1)";OPT
20020 SCREEN 1: COLOR 0,2:KEY OFF: CLS
20030 INC = .4
20040 LOCATE 1,2
20041 PRINT "PARABOLIC DISH  expert  novice"
20050 LOCATE  4,7 :PRINT"absolute"
20060 LOCATE  4,25: PRINT"relative"
20070 LOCATE 2,14:PRINT "    approximate "
20080 PSET(188,4),2:PSET(254,4),3
20081 PSET(248,12),1
20090 FACT=20:DT=.02
20100 C =.3 : G = 1: PED= SQR(2*C*G)
20110 IF OPT = 1 THEN GOSUB 20550
20120 IF OPT = 0 THEN GOSUB 20500
20130 IF OPT = 1 THEN GOSUB 20610
20140 EPS = .4: R =1.8: RE=R: RA = R
20150 PD=PED+EPS:RD=0:RAD=RD:RED=0:PAD=PD
20160 IF OPT=1 THEN P = 3.1415/2:PE=P:PA=P
20170 PDD= - 2* RD*PD/R
20180 PADD= - 2* RAD*PAD/RA
20190 RDD = (R*PD^2-2*C*R*(G+2*C*RD^2))
20191 RDD = RDD/(1+(2*C*R)^2)
20200 RADD = (RA*PAD^2-2*C*RA*G)
20210 PD = PDD*DT+PD
20220 PAD = PADD*DT+PAD
20230 RD=RDD*DT+RD
20240 RAD=RADD*DT+RAD
20250 P = PD*DT+P:R=R+DT*RD
20260 PE= PED*DT +PE
20270 PA= PAD*DT+PA:RA=RA+DT*RAD
20280 X = R*COS(P):Y=R*SIN(P)
20290 XE= RE*COS(PE):YE=RE*SIN(PE)
20300 XA =RA*COS(PA):YA=RA*SIN(PA)
20310 IF OPT = 1 THEN KK=3.1415/2 ELSE KK=0
```

```
20320 MX = R*COS(P-PE+KK):MY=R*SIN(P-PE+KK)
20330 MXE= RE*COS(PE-PE+KK)
20331 MYE=RE*SIN(PE-PE+KK)
20340 MXA= RA*COS(PA-PE+KK)
20341 MYA=RA*SIN(PA-PE+KK)
20350 IF OPT = 1 THEN GOTO 20430
20360 PSET( 80+FACT*X,100-FACT*Y),3
20370 PSET(220+FACT*MX,100-FACT*MY),3
20380 PSET( 80+FACT*XE,100-FACT*YE),2
20390 PSET( 80+FACT*XA,100-FACT*YA),1
20400 PSET(220+FACT*MXE,100-FACT*MYE),2
20410 PSET(220+FACT*MXA,100-FACT*MYA),1
20420 GOTO 20170
20430 XX=80+FACT*X
20431 YYA=199-COS(INC)*FACT*C*( X^2+ Y^2)
20432 YYB=-SIN(INC)*FACT* Y
20433 YY =YYA+YYB
20434 PSET(XX,YY) ,3
20435 XX=220+FACT*MX
20436 YYA = 199-COS(INC)*FACT*C*(MX^2+MY^2)
20437 YYB = -SIN(INC)*FACT*MY
20438 YY=YYA+YYB:PSET(XX,YY),3
20439 XX=80+FACT*XE
20440 YYA=199-COS(INC)*FACT*C*(XE^2+YE^2)
20441 YYB = -SIN(INC)*FACT*YE
20442 YY = YYA+YYB
20443 PSET(XX,YY) ,2
20444 XX=80+FACT* XA
20445 YYA=199-COS(INC)*FACT*C*( XA^2+ YA^2)
20446 YYB=-SIN(INC)*FACT* YA
20447 YY=YYA+YYB
20448 PSET(XX,YY),1
20450 XX=220+FACT*MXE
20451 YYA= 199-COS(INC)*FACT*C*(MXE^2+MYE^2)
20452 YYB = -SIN(INC)*FACT*MYE
```

```
20453 YY = YYA+YYB
20454 PSET(XX,YY),1
20460 XX=220+FACT*MXA
20461 YYA= 199-COS(INC)*FACT*C*(MXA^2+MYA^2)
20462 YYB = -SIN(INC)*FACT*MYA
20463 YY=YYA+YYB
20464 PSET(XX,YY),1
20490 GOTO 20170
20500 FOR I = 1 TO 3  ' plan view coords
20510 CIRCLE(80,100),I*FACT,1
20520 CIRCLE(220,100),I*FACT,1
20530 NEXT
20540 RETURN
20550 FOR I = 0 TO 1200 STEP 6
20560 X=(I-600)/600:Z=C*X^2*COS(INC)*3
20570 PSET(80 +3*FACT*X,199-3*FACT*Z),1
20580 PSET(220+3*FACT*X,199-3*FACT*Z),1
20590 NEXT
20600 RETURN
20610 FOR I = 1 TO 3
20615 YY=199-COS(INC)*FACT*C*I^2
20620 CIRCLE(80,YY),FACT*I,1,0,6.28,SIN(INC)
20630 CIRCLE(220,YY),FACT*I,1,0,6.28,SIN(INC)
20640 NEXT
20650 RETURN
```

CHAPTER V

Surfaces of revolution

5.1 HEMISPHERICAL AND PARABOLOIDAL DISHES COMPARED

The paraboloid is only one of a large family of possible surfaces of revolution about the z-axis, upon which a particle may slide freely with gravity acting downward parallel to the z axis. Only the paraboloid permits particles, independent of the radius r, to go around in circles with the same period of revolution. Thus a cloud of them at various values of r and ϕ can move in solid rotation, and give the appearance, in a steadily rotating frame, of being at rest. On the other hand, particles with no angular momentum, oscillate in an r, z plane (constant ϕ), but their periods depend upon the starting radius. So the paraboloid does not have isochronous motions in all directions, except approximately so, near the axis.

Experimentalists are attracted to a spherical surface (the hemisphere that is concave upward) because it can be effectively modeled simply by suspending the particle by an inextensible, massless wire from a point on the ceiling. The particle can swing freely in two directions. In this case neither purely azimuthal (constant r) nor purely radial (constant ϕ—and here we are referring to the horizontal coordinate r, not the radius of the sphere) oscillations have periods independent of the initial r. In a sense, therefore, the spherical surface (or spherical pendulum) is a worse way of trying to build platforms for reference frames than the paraboloid. It cannot support a population of experts in an equilibrium configuration of solid rotation. Still, it is interesting.

Appropriate dynamical equations for these systems can be obtained easily without elaborate coordinate transformations. The general form of the height z of the surface is

$$z = f(r). \tag{5.1}$$

The particle is acted upon by gravity g acting in the negative z direction, and by an (unknown) reactive force F that is normal to the surface. The potential energy per unit mass is gz. In the cylindrical coordinates r, ϕ, z in absolute space, the velocity components are $\dot{r}$, $r\dot{\phi}$, $\dot{z}$. They are at right angles, so the kinetic energy per unit mass is

$$\frac{1}{2}(\dot{r}^2 + r^2\dot{\phi}^2 + \dot{z}^2).$$

Because the reactive force is normal to the surface, and therefore can do no work on the particle sliding tangentially to it, the total energy cannot change with time

$$\frac{d}{dt}\left[\frac{1}{2}(\dot{r}^2 + r^2\dot{\phi}^2 + \dot{z}^2) + gz\right] = 0. \tag{5.2}$$

Neither gravity nor the reactive force can exert a torque on the particle in the azimuthal (ϕ) direction, so the angular momentum μ remains a constant, and we have

$$\mu \equiv r^2\dot{\phi}. \tag{5.3}$$

Equations (5.1), (5.2), and (5.3), with initial conditions and fixed angular momentum, completely determine the motion of the particle in time. By elimination they can be used to derive a differential equation for r with respect to time.

For the parabolic dish we choose $z = cr^2$, then $\dot{z} = 2cr\dot{r}$. We eliminate $\dot{\phi}$ from (5.2) by using (5.3) and obtain

$$\frac{d}{dt}\left[\frac{1}{2}\left(\dot{r}^2(1 + 4c^2r^2) + \frac{\mu^2}{r^2}\right) + gcr^2\right] = 0$$

or, performing the indicated differentiation,

$$\ddot{r}(1 + 4c^2r^2) = -2gcr + \frac{\mu^2}{r^3} - 4c^2r\dot{r}^2 \qquad (5.4)$$

that determines r as a function of time, and then we use 5.3 to compute ϕ. These expressions can be used for a numerical exploration of the *paraboloid of revolution*.

For a hemispherical dish (fig. 5.1) we write

$$(z - a)^2 = a^2 - r^2,$$

and going through the same steps obtain the expression

$$\frac{d}{dt}\left[\frac{1}{2}\left(\dot{r}^2\left(1 + \frac{r^2}{a^2 - r^2}\right) + \frac{\mu^2}{r^2}\right) + g(a - \sqrt{a^2 - r^2})\right] = 0.$$

Performing the indicated differentiation, we obtain

$$\ddot{r}\left(1 + \frac{r^2}{a^2 - r^2}\right) = -\frac{gr}{\sqrt{a^2 - r^2}} + \frac{\mu^2}{r^3} - \dot{r}^2\left(\frac{a^2r}{(a^2 - r^2)^2}\right) \qquad (5.5)$$

the equation for a particle sliding on a hemisphere, concave upward.

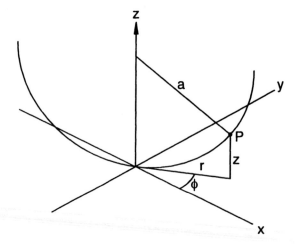

Figure 5.1. Coordinates for defining the hemispherical dish, concave up. The dish is of radius a. The coordinate system is cylindrical, and gravity is downward. The reactive force is directed along the radius a. In the figure the angle ϕ is measured in the plane $z = 0$.

5.2 COMPARISON WITH THE HOOKE SPRING PLANE

For comparison with the results of chapter 3, we can also write down the problem of simple harmonic motion on the slippery plane $z = 0$. Here, of course, there is no potential energy due to gravity, but there is potential energy per unit mass due to the Hooke spring of $k^2 r^2 / 2$.

Performing the same steps as in the other examples we have

$$\frac{d}{dt}\left[\frac{1}{2}\left(\dot{r}^2 + \frac{\mu^2}{r^2} + k^2 r^2\right)\right] = 0. \tag{5.6}$$

Performing the indicated differentiation we now have

$$\ddot{r} = \frac{\mu^2}{r^3} - k^2 r. \tag{5.7}$$

This is a somewhat unfamiliar form of the equation for *simple harmonic motion* on a plane with two degrees of freedom. In general the radius does not have a simple sinusoidal solution, unless the angular momentum vanishes. On the other hand r^2 is harmonic (section 5.5).

5.3 RESULTS FROM FIRST INTEGRALS

Let us use the first integrals as a means of exploring the phenomenology of the motion on the paraboloid and sphere in a little more detail, without referring to exotic functions. Equation (5.2) is a first integral of equation

(4.6) and expresses overall energy conservation. When combined with equation (5.3), the statement of the conservation of angular momentum, we obtain

$$\dot{z}^2 + \dot{r}^2 + \frac{\mu^2}{r^2} + 2gz = 2E/m, \tag{5.8}$$

where E is the total energy of the particle. It is convenient to introduce the equation for the surface $z = f(r)$ in an inverse form, namely

$$r^2 = G(z). \tag{5.9}$$

Taking one time derivative of equation (5.9) relates r and z in the form $2r\dot{r} = G'(z)\dot{z}$, which may be written as

$$\dot{r}^2 = \frac{G'^2(z)}{4G(z)}\dot{z}^2. \tag{5.10}$$

Substituting from (5.9) and (5.10) into equation (5.8) yields

$$\dot{z}^2\left[1 + \frac{G'^2(z)}{4G(z)}\right] + \frac{\mu^2}{G(z)} + 2gz = 2E/m,$$

or

$$\dot{z}^2 = \frac{\dfrac{2E}{m}G(z) - 2gzG(z) - \mu^2}{G(z) + \dfrac{1}{4}G'^2(z)}. \tag{5.11}$$

Equation (5.11) is a first order equation which gives $\dot{z}$ in terms of z. Of course a particle has the same speed as it crosses any level going up ($\dot{z} > 0$) as it has going down ($\dot{z} < 0$) through the same level.

Once we solve equation (5.11) for z as a function of time, equation (5.3), written in the form $\dot{\phi} = \mu/G(z)$, may be integrated for ϕ as a function of time. Now let us apply this to the paraboloid. We'll also pick up the simple harmonic motion—Hooke spring case—as an approximation to the paraboloidal case. Then we'll look at the spherical shell and try to compare these.

5.4 THE PARABOLOID

Let the paraboloid be represented in the following inverse form

$$r^2 = G_P(z) = 2az. \tag{5.12}$$

Then $G_P'(z) = 2a$ and $\frac{1}{4} G_P'^2(z) = a^2$. So equation (5.11) becomes

$$\dot{z}^2 = \frac{\dfrac{4E}{m}az - 4agz^2 - \mu^2}{a^2 + 2az}, \tag{5.13}$$

which we may write as

$$\dot{z}^2 = \frac{4ag(z - z_1)(z_2 - z)}{a^2 + 2az}. \tag{5.14}$$

The quantities z_1 and z_2 are the roots of the numerator on the right hand side of equation (5.13), and the choice of sign convention in equation (5.14) ensures that $\dot{z}^2$ is non-negative, and therefore z is real, for $z_1 < z < z_2$.

The two equations (5.13) and (5.14) for $\dot{z}^2$ must be identical, and therefore we see that $\mu^2 = 4agz_1z_2$ and $E/m = g(z_1 + z_2)$. Specifying z_1 and z_2 may be regarded as equivalent to setting the angular momentum and energy of the particle. Before actually integrating (5.14) there is one difficulty that we must address—the sign ambiguity in solving for $\dot{z} = \pm\sqrt{\dot{z}^2}$. A useful way around this difficulty is to represent z in terms of a parameter λ by writing $z = z_1 \cos^2\lambda + z_2 \sin^2\lambda$. As λ increases from 0 to $\pi/2$, z goes from z_1 to z_2 and as λ continues to increase, from $\pi/2$ to π, say, z goes from z_2 back to z_1. Thus the use of the parameter λ enables the oscillation to be represented in terms of a steadily increasing quantity.

To see how this works, note that $z - z_1 = (z_2 - z_1)\sin^2\lambda$ and $z_2 - z = (z_2 - z_1)\cos^2\lambda$, so

$$(z - z_1)(z_2 - z) = (z_2 - z_1)^2\sin^2\lambda\cos^2\lambda.$$

Also $\dot{z} = 2(z_2 - z_1)\sin\lambda\cos\lambda\,\dot{\lambda}$, so

$$\dot{z}^2 = 4\dot{\lambda}^2(z_2 - z_1)^2\cos^2\lambda\sin^2\lambda.$$

Substituting all of this into (5.14) simply leaves

$$\dot{\lambda}^2 = ag/(a^2 + 2az), \tag{5.15}$$

with z still given by

$$z = z_1\cos^2\lambda + z_2\sin^2\lambda. \tag{5.16}$$

Now the right hand side of (5.15) is positive, and we may take the positive square root on both sides without any ambiguity. The parameter λ will be an increasing function of time, and z will oscillate between z_1 and z_2. To complete the description of the motion we need to find ϕ, which we can do by equation (5.3). This is

$$\dot\phi = \frac{\mu}{r^2} = \frac{\mu}{2az} = \frac{\sqrt{agz_1z_2}}{az}. \tag{5.17}$$

So we simply solve (5.15) for $\dot\lambda$ and obtain

$$\dot\lambda = \sqrt{\frac{g}{a + 2z}}, \tag{5.18}$$

and then integrate (5.18) and (5.17) forward in time by some standard numerical scheme, with z given by (5.16) at each time step.

For the full paraboloidal problem with vertical acceleration kept in, the analytical integration of equation (5.15) leads to elliptic functions. The theory of these functions is too complicated for us to discuss here, but a good reference is Arthur Cayley's *An Elementary Treatise of Elliptic Functions* (1895, reprint: Dover, 1961). Beware, the term elementary in the title does not mean simple.

5.5 THE HOOKE SPRING PLANE

A simple example that can be carried out completely in analytical form is the Hooke spring problem. The mathematics is exactly the same as that governing the motion of a particle on a paraboloid, ignoring the vertical accelerations. This means that we omit the part of the kinetic energy due to the vertical component of velocity—that is, we drop $\dot z^2$ in equation (5.8) while retaining the $\dot r^2$ term.

We can still use z as a variable, where $r^2 = 2az$. Following through the steps we see that the only difference is that the $G(z)$ term is omitted in the denominator on the right hand side of equation (5.11). Thus in equations (5.13), (5.14), and (5.15), $a^2 + 2az$ is replaced by a^2. Then equation (5.15) becomes $\dot{\lambda} = \sqrt{g/a}$ and we find $\lambda = \sqrt{g/a} \cdot t$. Therefore

$$r^2 = 2az = 2az_1\cos^2\left(\sqrt{\frac{g}{a}}t\right) + 2az_2\sin^2\left(\sqrt{\frac{g}{a}}t\right).$$

Equation (5.17) becomes

$$\dot{\phi} = \frac{\mu}{r^2} = \frac{\sqrt{agz_1z_2}}{az_1\cos^2\left(\sqrt{\frac{g}{a}}t\right) + az_2\sin^2\left(\sqrt{\frac{g}{A}}t\right)},$$

or

$$\frac{d\phi}{d\lambda} = \frac{\sqrt{z_1z_2}}{z_1\cos^2\lambda + z_2\sin^2\lambda}. \tag{5.19}$$

A standard table of integrals gives the solution

$$\phi = \tan^{-1}\left(\sqrt{\frac{z_2}{z_1}}\tan\lambda\right),$$

which means

$$\tan^2\lambda = \frac{z_1}{z_2}\tan^2\phi,$$

or in turn,

$$\cos^2\lambda = \frac{1}{1 + \tan^2\lambda} = \frac{z_2\cos^2\phi}{z_1\sin^2\phi + z_2\cos^2\phi},$$

and

$$\sin^2\lambda = \frac{z_1\sin^2\phi}{z_1\sin^2\phi + z_2\cos^2\phi}.$$

Thus the equation $z = z_1 \cos^2\lambda + z_2 \sin^2\lambda$ becomes

$$z = z_1 z_2 / (z_1\sin^2\phi + z_2\cos^2\phi).$$

Since $z = r^2/2a$ this is $z_1 y^2 + z_2 x^2 = 2a z_1 z_2$ where we have written $x = r \cos \phi$ and $y = r \sin \phi$ to cast the answer in recognizable form. The orbit in absolute space is indeed an ellipse.

5.6 SPHERICAL DISH

We can now proceed to the spherical shell. Let its inverse equation be

$$r^2 = G(z) = 2az - z^2. \tag{5.20}$$

Then $\frac{1}{2} G'(z) = a - z$ and $G(z) + \frac{1}{4} G'^2(z) = a^2$. Thus, equation (5.11) becomes

$$\dot{z}^2 = \frac{\left(2\dfrac{E}{m} - 2gz\right)z(2a - z) - \mu^2}{a^2}. \tag{5.21}$$

Let us start by investigating graphically the zeros of the numerator of the right hand side of this equation. We set the numerator equal to zero, and write the resulting equation in the form $2g(E/mg - z) = \mu^2/(z(2a - z))$, and plot each side separately as a function of z (figure 5.2).

We see that when E is sufficiently large there are three roots which we may label z_1, z_2 and z_3. They satisfy

$$0 \leq z_1 \leq z_2 \leq 2a \leq z_3,$$

and also $\frac{1}{2}(z_1 + z_2) < a$. If the physical domain is not the entire interior surface of the sphere [e.g. a hemispherical cup] the possibility exists that there is only one root in the physical domain. In this latter case the particle will eventually escape from the constraint of the surface by over-shooting the rim. For the present let us consider situations in which z_1 and z_2 are within the physical domain, so the particle oscillates in the range $z_1 \leq z \leq z_2$. We write (5.21) in the form

$$\dot{z}^2 = \frac{2g}{a^2}(z - z_1)(z_2 - z)(z_3 - z), \qquad (5.22)$$

Making (5.21) and (5.22) identical implies

$$\mu^2 = 2gz_1z_2z_3; \qquad 2a\frac{E}{gm} = z_1z_2 + z_2z_3 + z_3z_1,$$

and

$$\frac{E}{gm} + 2a = z_1 + z_2 + z_3.$$

Eliminating E/gm between the last two equations gives

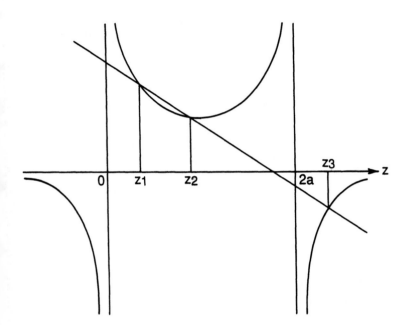

Figure 5.2. Location of turning latitudes (limiting altitudes) on the hemispherical dish.

$$z_3 = 2a + \frac{z_1 z_2}{2a - (z_1 + z_2)},$$

which means that we can pick z_1, z_2 and the radius a, but having done so, E, μ^2, and z_3 are all determined.

Introduce the parameter λ as before, so

$$z = z_1 \cos^2\lambda + z_2 \sin^2\lambda.$$

Substituting in (5.22) yields

$$\dot{\lambda}^2 = \frac{g}{2a^2}(z_3 - z), \qquad (5.23)$$

and the azimuth ϕ satisfies

$$\dot{\phi} = \frac{\mu}{2az - z^2}. \qquad (5.24)$$

These equations may be integrated by forward time stepping to obtain the orbit.

Consider the shape of the orbits qualitatively. One quantity of interest is the change in azimuth associated with changing z from z_1 to z_2. For the Hooke spring case we found from equation (5.19) that

$$\tan\phi = \sqrt{\frac{z_2}{z_1}}\tan\lambda,$$

so that as λ increases from 0 to $\pi/2$, so does ϕ. Therefore as z increases from z_1 to z_2, ϕ increases from 0 to $\pi/2$. Inasmuch as equation (5.19) can be written as

$$\frac{d\phi}{d\lambda} = \frac{\sqrt{z_1 z_2}}{z},$$

we see that for the Hooke spring problem ϕ increases more rapidly than λ on the inner portion of the orbit ($z < \sqrt{z_1 z_2}$) and less rapidly on the outer portion ($z > \sqrt{z_1 z_2}$), in just such a way that the two angles catch up with each other every half period of the inertial oscillation.

5.7 ROTATION OF THE APSIDES

Now consider two particles, one on a paraboloid and the other on a Hooke spring plane. If λ_P, ϕ_P refer to the paraboloid and λ_{HS}, ϕ_{HS} to the Hooke spring plane, we have

$$\dot{\lambda}_{HS}^2 = \frac{g}{a}, \qquad \dot{\phi}_{HS} = \frac{\mu}{2az}, \quad \text{and} \quad \dot{\lambda}_P^2 = \frac{g}{a+2z}, \qquad \dot{\phi}_P = \frac{\mu}{2az}.$$

For both cases

$$z = z_1 \cos^2\lambda + z_2 \sin^2\lambda$$

If both particles start at the same initial z and increase λ by a small amount $\Delta\lambda$, the times required to do this are

$$\Delta t_{HS} = \sqrt{\frac{a}{g}} \Delta\lambda$$

for the Hooke spring, and

$$\Delta t_P = \sqrt{\frac{a+2z}{g}} \Delta\lambda$$

for the paraboloid. The latter is longer. Since

$$\dot\phi_{HS} = \dot\phi_P = \mu/2az$$

is initially the same for both particles, we find that

$$\Delta\phi_P = \Delta t_P \cdot \frac{\Delta\phi_{HS}}{\Delta t_{HS}} > \Delta\phi_{HS}.$$

This holds for all z, so the change in azimuth on the paraboloid associated with a given change in z is always greater than the corresponding change in the Hooke spring plane. Therefore, on the paraboloid ϕ will increase by more than $\pi/2$ as z goes from z_1 to z_2. Hence, the total change in ϕ for a complete orbit (z_1 to z_2 and back twice) will exceed 2π. A point on the orbit closest to the center is called a lower apsis; one farthest from the center is called a higher apsis. From one orbit to the next the location of these apsides changes, so the apsides themselves rotate about the center.

When we experiment with the numerical exercises we note that the apsides not only rotate in the same direction as that of the platform in absolute space, but also rotate in the opposite direction in the rotating reference frame. This is the famous Rossby drift to the west that was first recognized in terrestrial atmospheric systems. To derive this result we need to be a little more mathematical, and for convenience we will treat the question nondimensionally. We choose length and time scales such that $g = \Omega = 1$. Then $a = 1$, and our equations are written in the form

$$z = z_1 \cos^2\lambda + z_2 \sin^2\lambda$$

$$\dot\lambda = \frac{1}{\sqrt{1+2z}}, \qquad \dot\phi = \frac{\sqrt{z_1 z_2}}{z} \equiv \frac{z_0}{z}.$$

The time T for a particle to move from z_1 to z_2 is that for which $\Delta\lambda = \pi/2$:

$$T = \int_0^{\pi/2} \frac{1}{\dot{\lambda}} d\lambda = \int_0^{\pi/2} \sqrt{1 + 2z} \, d\lambda > \frac{\pi}{2}. \qquad (5.26)$$

For simple harmonic motion it is $T = \pi/2$. The azimuthal angle Φ moved in this time is

$$\Phi = \int_0^{\pi/2} \frac{\dot{\phi}}{\dot{\lambda}} d\lambda = \int_0^{\pi/2} \frac{z_0}{z} \sqrt{1 + 2z} \, d\lambda. \qquad (5.27)$$

But from a standard table of integrals [note equation (5.19)] we know that

$$\int_0^{\pi/2} \frac{z_0}{z} d\lambda = \frac{\pi}{2}.$$

Therefore $\Phi > \pi/2$, or the line of apsides moves toward the east (in the sense of rotation of the system) in inertial space. The azimuthal move, with reference to the moving system Ωt is

$$\Phi - T = \int_0^{\pi/2} \left(\frac{z_0}{z} - 1\right) \sqrt{1 + 2z} \, d\lambda \qquad (5.28)$$

$$= \int_0^{\pi/2} \left(\frac{z_0}{z} - 1\right) (\sqrt{1 + 2z} - \sqrt{1 + 2z_0} + \sqrt{1 + 2z_0}) \, d\lambda.$$

But we know that

$$\int_0^{\pi/2} \left(\frac{z_0}{z} - 1\right) (\sqrt{1 + 2z_0}) \, d\lambda = \sqrt{1 + 2z_0} \left(\frac{\pi}{2} - \frac{\pi}{2}\right) = 0.$$

Therefore we have

$$\Phi - T = \int_0^{\pi/2} \left(\frac{z_0}{z} - 1 \right) [\sqrt{1 + 2z} - \sqrt{1 + 2z_0}] \, d\lambda$$

$$\equiv \int_0^{\pi/2} \mathcal{T} \, d\lambda.$$

Now inspection of this integral in two domains, $z < z_0$ and $z > z_0$ shows us that the integrand $\mathcal{T} < 0$ in each domain. When $z < z_0$, $(z_0/z - 1) > 0$, [ ] < 0, $\mathcal{T} < 0$; and when $z > z_0$, $(z_0/z - 1) < 0$, [ ] > 0, and $\mathcal{T} < 0$ again. Therefore $\Phi - T < 0$ and the drift of the apsides in the relative frame drift westward, counter to Ωt.

Much the same happens inside a spherical shell. It is possible, but not easy, to show from equations (5.22) and (5.23) that the change in azimuth to go from z_1 to z always lies between $\pi/2$ and π. (See, for example, chapter 15 in *A treatise on gyrostatics and rotational motion* by Andrew Gray.)

5.8 NUMERICAL SOLUTIONS

Although the Hooke plane has no surprises, to develop general analytical expressions for the motion of a particle under gravity in a paraboloid or hemisphere involves elliptic functions, and the phenomenology becomes complicated. Discussions of various degrees of completeness can be found in standard texts and treatises on classical particle mechanics. One encounters the hemisphere more often than the paraboloid.

Because of the formal complexities the numerical solutions are useful for exploring the phenomenology. Exercises 5-1 through 5-5 are computed in absolute space. The displays in Exercises 5-4 and 5-5 are made relative to a uniformly rotating expert by subtracting the instantaneous absolute azimuth of the expert from that of the novice, and plotting coordinates embodying the difference of azimuth. The plots are projections on the r, ϕ plane, so give no indication of z except in the elevation views of the surfaces shown at the bottom of the screen. Forming the relative trajectories in this fashion makes sense for the plane and paraboloid because on them all experts rotate about the axis at the same angular velocity, independent of radius, so we can think of them as defining a uniformly rotating "platform" upon which Coriolis forces are defined. Inside the hemisphere, however, the angular velocity of experts does depend upon radius, so there is no meaningful uniformly rotating platform for reference purposes.

Another thing to keep in mind is that the hemisphere has a finite radius, $r = a$, while the plane and paraboloid have radii that extend indefinitely outward. The top rim of the concave hemisphere can be spoken of as the "equator." In the classical spherical pendulum the pendulum bob is usually envisaged as supported by a massless rigid rod pivoted freely at the center of a sphere, so that the bob can move over the full surface of a complete sphere without falling off. The tension in the rod can change sign. When we consider a sliding particle it is safer to restrict the motion to the concave hemisphere, because otherwise it might fall off the surface of the upper half of the sphere in a free trajectory inside the volume of the sphere, and we would have to keep track of the reactive force of the surface on the particle to detect this possibility.

Here are some qualitative features of the trajectories that are interesting. The novice generally slides around inside the surfaces, avoiding the lowest point, between two limiting radii, where $\dot{r} = 0$. Unlike the plane case, the trajectory of the novice in the paraboloid is more complicated because its quasielliptical orbit tends to precess about the axis. Let us define an apse line as a radius that is perpendicular to the trajectory. The apse line on the Hooke plane, in absolute space, remains stationary. In the paraboloid the apse line tends to rotate in the same sense as the particle's rotation about the axis, but more slowly, so that in the moving reference system the rotation of the apse line is retrograde. There is no unique reference rotation rate for defining a reference frame for the hemisphere, so reference to moving experts is possible in different ways. In exercise 5-4 the reference expert is taken to be the same as for the other surfaces. In exercise 5-5 the reference expert for the hemispherical display is taken to be the expert who starts at the same starting point as the novice. The programs will beep if the radius of the hemisphere is exceeded, and phenomena observed subsequently should be interpreted with caution.

PROBLEMS

Problem 5.0. A particle slides, under vertical gravity, on the inside of an inverted conical surface of revolution, $z = \alpha r$. Find the energy equation and the equations of motion in absolute space. Write a program to exhibit the trajectory of a novice particle in absolute space. Is there a uniformly rotating population of experts that define a rotating reference frame? For given angular momentum are there limit circles? How does the apse line rotate?

Problem 5.1. Consider two particles. One is on the surface of a flat frictionless plane table, and the other particle, attached to the first by a massless string, hangs down from a hole in the center of the table. Describe the motion of this system.

Problem 5.2. Compare the motions in problems 5.0 and 5.1.

Problem 5.3. The Hooke spring orbit is the ellipse described in chapter 3 and illustrated in figure 3.1. Go back and identify the angles ϕ and λ in the picture. From the geometry of the figure find r and ϕ as functions of λ. Verify that $r^2 = 2az$ and that

$$\frac{d\phi}{d\lambda} = \frac{\sqrt{z_1 z_2}}{z} = \frac{r_1 r_2}{r^2}.$$

Problem 5.4. Write a computer program that computes motions of a particle by means of integrating the $\dot\lambda$ equations. You can use the graphic parts of exercise 5.1, so it ought to be fairly straightforward.

EXERCISES

Exercise 5-1 [lines 25000–25840] *Three surfaces of revolution.* This exercise shows plan view in absolute space of (1) the flat Hooke plane, (2) the paraboloid of revolution, (3) the hemisphere, concave up, from left to right. These numbers also correspond to the indices in the program for each surface. Below them there are also shown a vertical section through each surface. The subroutines for these plots are lines 25500–25750. The time step *DT,* and gravity *G* are in line 25030. The spring constant *K* is given in line 25040, and the corresponding constants for the other surfaces are computed in the same line so that all periods will be the same for small R. The angular momentum MU is the same for them all, in line 25050—you can change it later if you want to.

Initial conditions are set in lines 25060–25080. The initial default value of $R(I)$ is .5, and for $RD(I)$ is 0. These can also be changed. The first step of integrations of the equations, by simple forward differencing are carried out in lines 25090–25110. The next steps of integration are carried out in lines 25140–25150. The conversions to $X(I)$, $Y(I)$ and plotting are done in lines 25160–25170. Line 25180 returns to 25090 for another step of integration. The radius circumscribed on each of the panels is $R = 0.6$. You may choose to set this at 0.5 or 1.0 or to draw more index circles to mark the radial scale.

```
25000 'Exercise 5-1: Three surfaces of revol**
25010 ' Indices (1) simple harmonic motion
25011 '(2)paraboloid
25012 '(3)concave up hemisphere
25020 SCREEN 1: COLOR 0,2: KEY OFF:CLS
25030 DT = .04: G = 1: D =1
25040 K=1: K2=K^2: C=K2/2: A=K2/G
25050 MU = .2
25055 GOSUB 25500:GOSUB 25600
25060 FOR I = 1 TO 3:R(I)=.5: NEXT
25070 FOR I = 1 TO 3:RD(I)=0: NEXT
25090 RDD(1)=MU^2/(R(1)^3)-K^2*R(1)
25091 AAA=(1+(2*C*R(2))^2)
25092 CCC=MU^2/(R(2)^3)
25093 BBB=-2*G*C*R(2)-R(2)*(2*C*RD(2))^2
25100 RDD(2)=(BBB+CCC)/AAA
25105 AA= -G*R(3)/(SQR(A^2-R(3)^2))
25106 CC= - RD(3)^2*(A^2*R(3)/((A^2-R(3)^2)^2))
25107 DD = 1+R(3)^2/(A^2-R(3)^2)
25108 BB=  MU^2/(R(3)^3)
25110  RDD(3)=(AA+BB+CC)/DD
25140 FOR I = 1 TO 3: RD(I)=RDD(I)*DT+RD(I)
25141 R(I)=RD(I)*DT+R(I):NEXT
25150 FOR I = 1 TO 3:  PD(I)=MU/(R(I)^2)
25151 P(I)=PD(I)*DT+P(I)   :NEXT
25160 FOR I = 1 TO 3: X(I)=R(I)*COS(P(I))
25161 Y(I)=R(I)*SIN(P(I))  :NEXT
25170 FOR I = 1 TO 3
25171 PSET (50*X(I)+70+80*(I-1),60-50*Y(I)),I
25172 NEXT
25180 GOTO 25090
25500 FOR I = 0 TO 2: CIRCLE (70+80*I,60),30,1
25501 NEXT
25505 LOCATE 23,5
25506 PRINT "  plane   parabol   sphere";
```

```
25507 LOCATE 1,6
25508 PRINT "Three surfaces of revolution ";
25510 RETURN
25600 FOR I =-.7 TO .7 STEP .02:'side view
25610 L(1)=50*I +70+80*0: M(1)=170
25620 L(2)=50*I +70+80*1: M(2)=170-50*C*I^2
25625 IF (A^2-I^2)=<0 THEN GOTO 25700
25630 L(3)=50*I +70+80*2
25631 M(3)=170 -50*( A -SQR(A^2-I^2) )
25700 PSET(L(1),M(1)),1
25701 PSET(L(2),M(2)),1: PSET (L(3),M(3)),1
25725 NEXT
25750 RETURN
```

Exercise 5-2 [lines 26000–26440] *Oscillations in a vertical plane on the three platforms.* If you are typing in this program it will be helpful to start with exercise 5-1 and write the new lines over it. The main difference is that the angular momentum *MU* is taken as zero, so that the particles oscillate through the center within a vertical plane with $P(i) = 0$. There is a trap in lines 26270–26280 that prints out two times the time elapsed from the initial release of the particles at $R(i) = .5$ to the moment at the extreme of swing when the radial velocity changes sign. The value of *DT* is set small so that some accuracy is obtained, and when the whole swing (half period) is complete, the period is displayed for each case.

```
26000 'Exercise 5-2: 3 surf rev PLANE PERIODS*
26010 ' Indices (1) simple harmonic motion
26011 ' Indices (2) paraboloid
26012 ' Indices (3) concave up hemisphere
26020 SCREEN 1: COLOR 0,2: KEY OFF:CLS
26030 DT = .01: G = 1: D =1
26040 K=1: K2=K^2: C=K2/2: A=K2/G
26050 MU = 0
26060 GOSUB 26320:GOSUB 26370
26070 FOR I = 1 TO 3:R(I)=.5: NEXT
26080 FOR I = 1 TO 3:RD(I)=0: NEXT
26100 RDD(1)=MU^2/(R(1)^3)-K^2*R(1)
26105 AAA= (1+(2*C*R(2))^2)
26106 BBB= -2*G*C*R(2)-R(2)*(2*C*RD(2))^2
26107 CCC= MU^2/(R(2)^3)
26110 RDD(2)=(BBB+CCC)/AAA
26120 AA= -G*R(3)/SQR((A^2-R(3)^2))
26121 BB=MU^2/(R(3)^3)
26130 CC=-RD(3)^2*(A^2*R(3)/((A^2-R(3)^2)^2))
26140 DD = 1+R(3)^2/(A^2-R(3)^2)
26150  RDD(3)=(AA+BB+CC)/DD
26220 FOR I = 1 TO 3: RD(I)=RDD(I)*DT+RD(I)
26221 R(I)=RD(I)*DT+R(I):NEXT
26230 FOR I = 1 TO 3:  PD(I)=MU/(R(I)^2)
26231 P(I)=PD(I)*DT+P(I)   :NEXT
26240 FOR I = 1 TO 3: X(I)=R(I)*COS(P(I))
26241 Y(I)=R(I)*SIN(P(I))  :NEXT
26250 FOR I = 1 TO 3
26251 PSET (50*X(I)+70+80*(I-1),60-50*Y(I)),I
26252 NEXT
26260 FOR I = 1 TO 3 : LOCATE 15+I,5
26261 IF FLAG(I)=1 THEN GOTO 26280
26265 ZZZ=RD(I)*RDL(I)
26270 IF ZZZ<0 THEN GOTO 26271 ELSE GOTO 26280
26271 FLAG(I)=1
```

```
26272 PRINT USING "t(#)= ##.##";I,2*T
26280 RDL(I)=RD(I)
26290 NEXT
26300 T = T+DT
26310 GOTO 26100
26320 FOR I = 0 TO 2
26321 CIRCLE (70+80*I,60),30,1: NEXT
26330 LOCATE 23,5
26331 PRINT " plane    parabol  sphere";
26340 LOCATE 1,5
26341 PRINT " Three surfaces of revolution";
26350 LOCATE 2,5
26351 PRINT "PERIODS FOR PLANE OSCILLATIONS";
26360 RETURN
26370 FOR I =-.7 TO .7 STEP .02:'side view
26380 L(1)=50*I +70+80*0: M(1)=170
26390 L(2)=50*I +70+80*1: M(2)=170-50*C*I^2
26400 IF (A^2-I^2)=<0 THEN GOTO 26420
26410 L(3)=50*I +70+80*2
26411 M(3)=170 -50*( A -SQR(A^2-I^2) )
26420 PSET(L(1),M(1)),1: PSET(L(2),M(2)),1
26421 PSET (L(3),M(3)),1
26430 NEXT
26440 RETURN
```

Exercise 5-3 [lines 27000–27440] *Zonal periods for the three surfaces.* The angular velocities *PD(i)* are all set so that at the radius *R(I)* = .5 the orbits will be pure circles in the absolute reference frame [27100–27130]. The program runs until an intercept [27270] detects that *P(I)* is twice *pi* and then prints out the period. This could be done analytically without integration, of course, because the *PD(I)* remain constant.

```
27000 'Exercise 5-3: 3 surfaces of revolution
27001 ' ZONAL PERIODS ************
27010 ' Indices (1) simple harmonic motion
27011 '(2) paraboloid(3) concave up hemisphere
27020 SCREEN 1: COLOR 0,2: KEY OFF:CLS
27030 DT = .01: G = 1: D =1
27040 K=1: K2=K^2: C=K2/2: A=K2/G:B=(2*K*D)^2/G
27041 K=1: K2=K^2: C=K2/2: A=K2/G:B=(2*K*D)^2/G
27060 GOSUB 27320:GOSUB 27370
27070 FOR I = 1 TO 3:R(I)=.5: NEXT
27100 PD(1)=K
27110 PD(2)=SQR(2*G*C)
27120 PD(3)=SQR(G/(SQR(A^2-R(3)^2)))
27140 FOR I = 1 TO  3
27150 P(I)=PD(I)*DT+P(I)
27160 NEXT
27240 FOR I = 1 TO 3: X(I)=R(I)*COS(P(I))
27241 Y(I)=R(I)*SIN(P(I))  :NEXT
27250 FOR I = 1 TO 3
27251 PSET (50*X(I)+70+80*(I-1),60-50*Y(I)),I
27252 NEXT:NP=6.283
27260 FOR I = 1 TO 3 : LOCATE 15+I,5
27261 IF FLAG(I)=1 THEN GOTO 27290
27270 IF P(I)>NP GOTO 27271 ELSE GOTO 27290
27271 PRINT USING "t(#)= ##.##";I,2*T :FLAG(I)=1
27290 NEXT
27300 T = T+DT
27310 GOTO 27100
27320 FOR I = 0 TO 2
27321 CIRCLE (70+80*I,60),30,1: NEXT
27330 LOCATE 23,5
27331 PRINT " plane   parabol  sphere";
27340 LOCATE 1,7
27341 PRINT "Three surfaces of revolution";
27350 LOCATE 2,5
```

```
27351 PRINT "PERIODS FOR ZONAL OSCILLATIONS";
27360 RETURN
27365 'subroutine for side view
27370 FOR I =-.7 TO .7 STEP .02
27380 L(1)=50*I +70+80*0: M(1)=170
27390 L(2)=50*I +70+80*1: M(2)=170-50*C*I^2
27400 IF (A^2-I^2)=<0 THEN GOTO 27420
27410 L(3)=50*I +70+80*2
27411 M(3)=170 -50*( A -SQR(A^2-I^2) )
27420 PSET(L(1),M(1)),1
27421 PSET(L(2),M(2)),1: PSET (L(3),M(3)),1
27430 NEXT
27440 RETURN
```

Exercise 5-4 [lines 28000−28390] *Exercise 5-1 in the rotating reference frame of the plane example.* This program can be written over 5-1 with very few line changes. Note line 28210, which defines the angle of platform rotation, $P(0)$ and the line 28240, which inserts this angle into the argument of the expressions for the plotting variables $X(I)$, $Y(I)$.

Perhaps the most interesting thing to vary in this program is the angular momentum MU [line 28050]. There are significant differences in the relative orbits for $I = 2, 3$.

```
28000 'Exercise 5-4: Three surfaces of
28001 'revolution  RELATIVE TO ROTATING PLANE
28010 ' Indices (1) simple harmonic motion
28011 ' (2)paraboloid(3)concave up hemisphere
28020 SCREEN 1: COLOR 0,2: KEY OFF:CLS
28030 DT = .04: G = 1: D =1
28040 K=1: K2=K^2: C=K2/2: A=K2/G
28050 MU = .2
28060 GOSUB 28270:GOSUB 28320
28070 FOR I = 1 TO 3:R(I)=.5: NEXT
28080 FOR I = 1 TO 3:RD(I)=0: NEXT
28100 RDD(1)=MU^2/(R(1)^3)-K^2*R(1)
28105 AAA=1+(2*C*R(2))^2
28106 CCC= -2*G*C*R(2)-R(2)*(2*C*RD(2))^2
28107 BBB= MU^2/(R(2)^3)
28110 RDD(2)=(BBB+CCC)/AAA
28120 AA= -G*R(3)/(SQR(A^2-R(3)^2))
28121 BB=   MU^2/(R(3)^3)
28130 CC= - RD(3)^2*(A^2*R(3)/((A^2-R(3)^2)^2))
28140 DD = 1+R(3)^2/(A^2-R(3)^2)
28150  RDD(3)=(AA+BB+CC)/DD
28210 P(0)=K*DT+P(0)
28220 FOR I = 1 TO 3: RD(I)=RDD(I)*DT+RD(I)
28221 R(I)=RD(I)*DT+R(I):NEXT
28230 FOR I = 1 TO 3:  PD(I)=MU/(R(I)^2)
28231 P(I)=PD(I)*DT+P(I)   :NEXT
28240 FOR I = 1 TO 3: X(I)=R(I)*COS(P(I)-P(0))
28241 Y(I)=R(I)*SIN(P(I)-P(0)):NEXT
28250 FOR I = 1 TO 3
28251 AAA= 50*X(I)+70+80*(I-1)
28252 BBB= 60-50*Y(I)
28253 PSET (AAA,BBB),1:NEXT
28255 T= T+DT: XLS=XS: YLS=YS
28256 XS = 150+50*COS(T):YS=150+50*SIN(-T)
28260 GOTO 28100
```

```
28270 FOR I = 0 TO 2
28271 CIRCLE (70+80*I,60),30,1: NEXT
28280 LOCATE 23,5
28281 PRINT " plane    parabol   sphere";
28290 LOCATE 1,5
28291 PRINT " 3 surfaces of revolution ";
28300 LOCATE 2,5
28301 PRINT " RELATIVE TO ROTATING PLANE";
28310 RETURN
28320 'subroutine for side view
28321 FOR I =-.7 TO .7 STEP .02
28330 L(1)=50*I +70+80*0: M(1)=170
28340 L(2)=50*I +70+80*1: M(2)=170-50*C*I^2
28350 IF (A^2-I^2)=<0 THEN GOTO 28370
28360 L(3)=50*I +70+80*2
28361 M(3)=170 -50*( A -SQR(A^2-I^2) )
28370 PSET(L(1),M(1)),1
28371 PSET(L(2),M(2)),1: PSET (L(3),M(3)),1
28380 NEXT
28390 RETURN
```

Exercise 5-5 [lines 29000–29370] *Relative to local rotating experts.* When one examines the display in exercise 5-4 it seems odd, at first, to see the novice particle in the relative system of the hemisphere drifting sometimes in the positive (counter-clockwise) direction, while the novice in the paraboloid always drifts clockwise. However, upon reflection, it will become clear that the hemisphere differs from the paraboloid in that it is not a platform upon which experts can rotate around the axis with the same angular velocity regardless of radius. In fact an expert at the very rim of the hemisphere ($r = a$) would have to ride around on a vertical surface of infinite slope, and so travel at infinite speed. The expert's angular velocity for the parabola is given by $\dot{\phi}_e = \sqrt{2cg}$ whereas the expert on the hemisphere rotates about the axis at a speed $\dot{\phi}_e = (a^2 - r^2)^{-1/4}g^{1/2}$. Unless $r << a$ the expert on the hemisphere therefore must travel much more quickly than that on either the Hooke plane or the paraboloid. In our display in exercise 5-4, we simply subtracted the angular velocity of an expert on the plane or paraboloid from each of the instantaneous angles computed in absolute space (of exercise 5-1), so the novice in the hemisphere really does not have its own expert's angular velocity subtracted from it to show its trajectory relative to an expert at its starting radius. Here in exercise 5-5 we have remedied this, as can be seen by noting the general definition of $R(I)$ in line 29040, and the new definition of the angular velocity of the hemisphere's expert $P(5)$ in line 29150, and the recalculation of the $X(3)$ and $Y(3)$ just before plotting in line 29190. This assures that we subtract a different angle from the instantaneous angle in absolute space when plotting the novice in the hemisphere [29200]. However, we did not change the angular momentum MU. It is rather interesting to try larger values of *MU* [line 29050] and larger initial radii *RE* (29040], so that we can see what happens when the novices approach $R = A$. In this program A happens to be unity. The radius circumscribed in the display is at $R = .6$. You will notice that because of the steep lip of the hemisphere the novice cannot pass $R = A$, but can easily do so on the other two surfaces. This makes a big qualitative difference in the trajectories.

```
29000 'Exercise 5-5: 3 surfaces of revolution
29001 ' RELATIVE TO ROTATING EXPERTS"
29010 ' Indices (1) simple harmonic motion
29011 '(2)paraboloid,(3)concave up hemisphere
29020 SCREEN 1: COLOR 0,2: KEY OFF:CLS
29030 DT = .04: G = 1: D =1
29040 K=1: K2=K^2: C=K2/2: A=K2/G : RE=.5
29050 MU = .2
29060 GOSUB 29240:GOSUB 29300
29070 FOR I = 1 TO 3:R(I)=RE: NEXT
29080 FOR I = 1 TO 3:RD(I)=0: NEXT
29090 RDD(1)=MU^2/(R(1)^3)-K^2*R(1)
29091 AAA=1+(2*C*R(2))^2
29092 CCC= -2*G*C*R(2)-R(2)*(2*C*RD(2))^2
29093 BBB= MU^2/(R(2)^3)
29100 RDD(2)=(BBB+CCC)/AAA
29110 AA= -G*R(3)/(SQR(A^2-R(3)^2))
29111 BB=  MU^2/(R(3)^3)
29120 CC=-RD(3)^2*(A^2*R(3)/((A^2-R(3)^2)^2))
29130 DD = 1+R(3)^2/(A^2-R(3)^2)
29140  RDD(3)=(AA+BB+CC)/DD
29150 P(0)=K*DT+P(0)
29151 P(5)=(A^2-RE^2)^(-1/4)*DT+P(5)
29160 FOR I = 1 TO 3: RD(I)=RDD(I)*DT+RD(I)
29161 R(I)=RD(I)*DT+R(I):NEXT
29170 FOR I = 1 TO 3:  PD(I)=MU/(R(I)^2)
29171 P(I)=PD(I)*DT+P(I)   :NEXT
29180 FOR I = 1 TO 3: X(I)=R(I)*COS(P(I)-P(0))
29181 Y(I)=R(I)*SIN(P(I)-P(0)):NEXT
29190 X(3)=R(3)*COS(P(3)-P(5))
29191 Y(3)=R(3)*SIN(P(3)-P(5))
29200 FOR I = 1 TO 3
29201 PSET (50*X(I)+70+80*(I-1),60-50*Y(I)),I
29202 NEXT
29210 T= T+DT: XLS=XS: YLS=YS
```

```
29220 XS = 150+50*COS(T):YS=150+50*SIN(-T)
29230 GOTO 29090
29240 FOR I=0 TO 2:CIRCLE(70+80*I,60),30,1
29241 NEXT
29250 LOCATE 23,5
29251 PRINT " plane    parabol   sphere";
29260 LOCATE 1,5
29261 PRINT " Three surfaces of revolution ";
29270 LOCATE 2,7
29271 PRINT "RELATIVE TO ROTATING EXPERTS";
29280 LOCATE 3,8: PRINT "all with same mu";
29290 RETURN
29300 FOR I=-.9 TO .9 STEP .02:'sub sideview
29310 L(1)=50*I +70+80*0: M(1)=170
29320 L(2)=50*I +70+80*1: M(2)=170-50*C*I^2
29330 IF (A^2-I^2)=<0 THEN GOTO 29350
29340 L(3)=50*I +70+80*2
29341 M(3)=170 -50*( A -SQR(A^2-I^2) )
29350 PSET(L(1),M(1)),1: PSET(L(2),M(2)),1
29351 PSET (L(3),M(3)),1
29360 NEXT
29370 RETURN
```

Exercise 5-6 [lines 24000−24300] *Particle on a paraboloid computed by first integrals.* This program computes the trajectory of a particle on a paraboloidal surface from first integrals. After setting up some graphics [24020−24070] default values of the parameters are defined [24080−24090], then more graphics for the display coordinates [24100−24150]. An input of the two limiting altitudes z_1 and z_2 is then asked for [24160]. These should be between 0 and 1 to ensure that they keep within the limits of the display.

The integration loop then begins [24170] where we first compute z from L (L = lambda). The suffix D stands for time derivative, *dot,* and the suffix H means computed without vertical acceleration, as in the Hooke plane. P is the azimuthal angle ϕ. Because the display on the left panel is in a ϕ, r plane there is a reset of P as it runs offscale [24230]. The three curves as drawn on the left panel are: [24250], the trajectory in inertial space; [24270], the trajectory in the rotating frame; and [24260], the corresponding trajectory in the Hooke plane. Line 24280 draws, on the right panel, the difference between the true azimuth and the Hooke-plane azimuth (note that Q is P with a different starting point). The right hand panel shows the gradual westward (negative) drift of the relative trajectory with respect to a Hooke trajectory. The loop is completed by [24290−24300].

```
24000 '5-6 Paraboloid from first integrals
24010 SCREEN 1: COLOR 0,2: KEY OFF: CLS
24020 LOCATE 2,25:PRINT "^"
24030 LOCATE 3,25:PRINT "r"
24040 LOCATE 24,1:PRINT "< p  ";
24050 LOCATE 24,25:PRINT "0";
24060 LOCATE 14,35:PRINT "p-ph";
24070 LINE (200,100)-(320,100),1
24080 DT =.1:PI=4*ATN(1):P = PI:Q=PI :PH=PI
24090 A=1:G=1:OMEGA = SQR(G/A)
24100 FOR I = 0 TO 2 STEP .02
24110 Y=I^2/(2*A)
24120 PSET(30*2*3.1415-100*I,180-100*Y),1
24130 NEXT
24140 FOR I = 0 TO 2 STEP 1
24150 LINE(0,180)-(30*2*3.1415,180-100*I),1,B
24160 NEXT:LOCATE 1,1:INPUT "z1,z2";Z1,Z2
24170 Z = Z1*(COS(L))^2+Z2*(SIN(L))^2
24171 LH=SQR(G/A)*T
24180 ZH= Z1*(COS(LH))^2+Z2*(SIN(LH))^2
24190 RH=SQR(2*A*ZH)
24200 LD= SQR(G/(A+2*Z))
24210 PD=(SQR(A*G*Z1*Z2))/(A*Z)
24211 PHD=(SQR(A*G*Z1*Z2))/(RH^2/2)
24220 L = LD*DT+L: P=PD*DT+P
24221 Q=PD*DT+Q :PH=PHD*DT+PH
24230 IF P > 2*PI THEN P =P -2*PI
24240 R=SQR(2*A*Z):PR=Q-OMEGA*T
24250 PSET(30*(2*PI-P),180-100*R),3
24260 PSET(30*(2*PI-PH+OMEGA*T),180-100*RH),1
24270 PSET(30*(2*PI-PR),180-100*R),2
24280 PSET(200+3*T,100-10*(Q-PH)),3
24290 T = T+DT
24300 GOTO 24170
```

Exercise 5-7 [lines 24500−24730] *Oblique view of particle sliding on paraboloidal dish, by first integrals.* The inclination of the dish is set in line 24530, the default being *PI/7*. The view of the dish is obtained by drawing circles of equal radius, at the heights required to fall upon the paraboloid, and projecting them on an oblique viewing plane [24540−24580]. After input of the two limiting heights [24590] the integration proceeds as in the previous example [24600−24650], then rectangular coordinates are computed in the two reference frames, inertial and rotating, [24660−24670]. These are rotated for the projection and plotted [24675−24690] on the projected plane. Color 2 is the trajectory in resting inertial space, Color 3 is the trajectory in the rotating reference frame. Color 2 is changed to Color 5 when the particle is behind the rim of the dish as defined by the height of outermost radius [24687]. Time is stepped up by the increment *DT* [24720] and the integration loop reentered [24730]. With default values and choice of $Z1 = 0.1$, $Z2 = 0.4$ one gets a good view of the retrograde rotation of the line apsides in the rotating frame.

```
24500 '5-7  Paraboloid from first integrals.
24501 'Oblique view of the dish
24510 SCREEN 1: COLOR 0,2: KEY OFF: CLS
24520 DT =.05 :PI=4*ATN(1)  :P=PI/2
24530 INC = PI/7
24540 A=3 :G=1:OMEGA=SQR(G/A)
24550 FOR  HT = 0 TO 1 STEP .2
24560 RAD = SQR(2*A*HT)
24561 YY = 100-50*HT*COS(INC)
24570 CIRCLE(150,YY),50*RAD,1,,,SIN(INC)
24580 NEXT
24590 LOCATE 24,1:INPUT"z1,z2";Z1,Z2
24600 Z = Z1*(COS(L))^2+Z2*(SIN(L))^2
24610 LD= SQR(G/(A+2*Z))
24620 PD=(SQR(A*G*Z1*Z2))/(A*Z)
24630 L = LD*DT+L: P=PD*DT+P
24640 R=SQR(2*A*Z)
24650 PR=P-OMEGA*T
24660 X=R*COS(P):Y=R*SIN(P):Z=R^2/(2*A)
24670 XR=R*COS(PR):YR=R*SIN(PR):Z=R^2/(2*A)
24675 YYY=100-50*YR*SIN(INC)-50*Z*COS(INC)
24680 PSET(150+50 *XR,YYY),3
24684 RADT=SQR(2*A):SQT=SQR(1-X^2/RADT^2)
24685 ZZ=    COS(INC)- RADT*SIN(INC)*SQT
24686 YY=Y*SIN(INC)+Z*COS(INC)
24687 IF YY<ZZ THEN CLR = 5 ELSE CLR = 2
24688 YYY=100-50*Y*SIN(INC)-50*Z*COS(INC)
24690 PSET(150+50 *X,YYY),CLR
24710 RO=R:ZO=Z
24720 T = T+DT:IF P > 2*3.14159 THEN P = 0
24730 GOTO 24600
```

CHAPTER VI

Velocity and acceleration in spherical coordinates

6.1 TRANSFORMATION FROM CYLINDRICAL POLAR COORDINATES TO SPHERICAL COORDINATES

For application to terrestrial phenomena it is natural to use spherical coordinates. In figure 6.1 the center of a sphere is denoted by O, and the polar axis by AA'. The point P has three coordinates: R, the radial distance from the center O, ϑ the angle (latitude) between the radius OP and the equatorial plane EOS, and ϕ the angle (longitude) measured eastward from the meridional plane $ASA'O$, fixed in celestial space, with S representing some celestial point, say the longitude of Aries.

The velocities at P are u (eastward), v (northward), and w (vertically upward), as shown in figure 6.2. They are mutually perpendicular. As we anticipate from the fact that the expressions for radial and tangential acceleration in a plane [equations (2.6)] are more complex than for the rectilinear accelerations, the expressions for the three components of acceleration in spherical coordinates are even more complicated. They can be obtained by vectorial manipulation, as available in standard texts (e.g. Morse and Feshbach 1957 or Batchelor 1967), but a geometrical derivation is revealing physically. In searching for one in the literature we found the graphical derivations given in

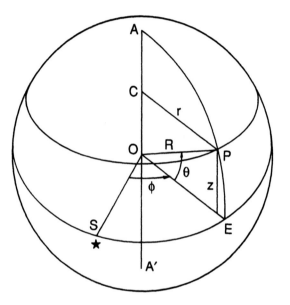

Figure 6.1. The position of the point P is defined in two coordinate systems: a cylindrical coordinate system r, ϕ, z; a spherical coordinate system ϑ, ϕ, R. The coordinate ϕ is common to both of them. The line AA' defines the polar axis. ES is the equator. Absolute longitude ϕ is measured from point S, fixed with respect to the stars. The center of the sphere is at O; the center of the small latitude circle through P is at C.

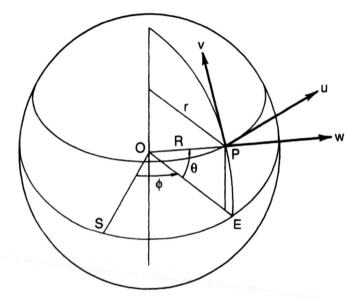

Figure 6.2. Orientation of spherical coordinate velocity components u, east, v, north, and w upwards, on the sphere.

the texts by D. Brunt and by B. Haurwitz obscure, if not flawed. So we offer it in another form which seems clearer to us.

The idea will be to start with cylindrical polar coordinates where r is the distance of the point P from the point C on the axis of rotation, and ϕ the longitude. The third coordinate will be the distance z of P from the equatorial plane. From our knowledge of the equations in polar form, we can find the two components of acceleration at P in the tangential direction (eastward), and outward in the plane of the small latitude circle with center C:

$$A_\phi = r\ddot{\phi} + 2\dot{r}\dot{\phi}, \tag{6.1}$$

and

$$A_r = \ddot{r} - r\dot{\phi}^2. \tag{6.2}$$

The acceleration parallel to the axis is simply

$$A_z = \ddot{z}. \tag{6.3}$$

Now, because by the geometry

$$z = R\sin\vartheta, \qquad r = R\cos\vartheta, \tag{6.4}$$

and the corresponding time derivatives are

$$\dot{z} = \dot{R}\sin\vartheta + R\cos\vartheta\,\dot{\vartheta},$$
$$\ddot{z} = \ddot{R}\sin\vartheta + 2\dot{R}\cos\vartheta\,\dot{\vartheta} - R\sin\vartheta\,\dot{\vartheta}^2 + R\cos\vartheta\,\ddot{\vartheta},$$
$$\dot{r} = \dot{R}\cos\vartheta - R\sin\vartheta\,\dot{\vartheta},$$

and

$$\ddot{r} = \ddot{R}\cos\vartheta - 2\dot{R}\sin\vartheta\,\dot{\vartheta} - R\cos\vartheta\,\dot{\vartheta}^2 - R\sin\vartheta\,\ddot{\vartheta}. \tag{6.5}$$

We can write the above cylindrical components of acceleration in terms of the spherical coordinates r, ϑ, ϕ.

$$A_\phi = R\cos\vartheta\,\ddot\phi + 2\dot\phi(\dot R\cos\vartheta - R\sin\vartheta\,\dot\vartheta),$$

$$A_z = \ddot R\sin\vartheta + 2\dot R\cos\vartheta\,\dot\vartheta - R\sin\vartheta\,\dot\vartheta^2 + R\cos\vartheta\,\ddot\vartheta,$$

and

$$A_r = \ddot R\cos\vartheta - 2\dot R\sin\vartheta\,\dot\vartheta - R\cos\vartheta\,\dot\vartheta^2$$
$$- R\sin\vartheta\,\ddot\vartheta - R\cos\vartheta\,\dot\phi^2. \tag{6.6}$$

However the components of acceleration in the spherical coordinates' directions now must be obtained by projecting the components in the cylindrical coordinates r, ϕ, z upon the correct directions. The two sets of directions are shown in figures 6.3 and 6.4. By study of these figures it is evident that the eastward direction of acceleration A_E is the same in both coordinate systems. So we can immediately write down the expression for acceleration in this direction in spherical coordinates. This is

$$A_E = R\cos\vartheta\,\ddot\phi + 2\dot\phi(\dot R\cos\vartheta - R\sin\vartheta\,\dot\vartheta). \tag{6.7}$$

On the other hand we can see that a projection is going to be necessary to transform the accelerations in the APO meridional plane from one set to the other. In figure 6.5 we have drawn the various vectors in the plane APO. We have denoted the acceleration along CP by A_r and the z acceleration by A_z. Then if we define the northward acceleration in spherical coordinates by A_N and the vertical one (outward along OP) by A_V, these must be related by the following composition of components

$$A_N = A_z\cos\vartheta - A_r\sin\vartheta, \tag{6.8}$$

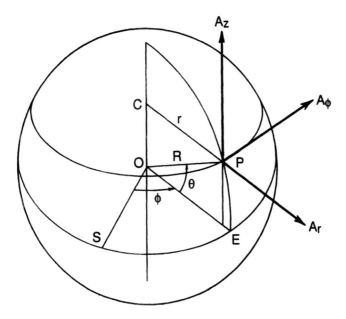

Figure 6.3. The orientation of the cylindrical coordinate components of the acceleration (in the r, ϕ, z system).

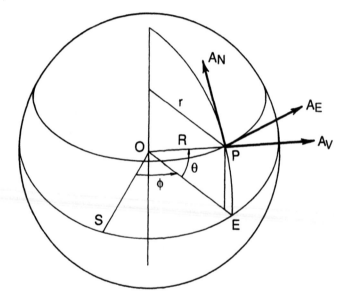

Figure 6.4. The orientation of the spherical coordinate components of the acceleration (A_E is east, A_N is north, and A_V is upwards). We tend to think geographically about this coordinate system, but remember we are not thinking about a rotating earth yet.

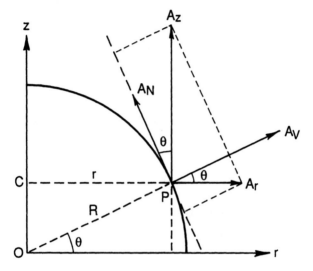

Figure 6.5. This diagram is drawn in the r, z plane, in a meridional plane ($AOEP$ of figure 6.1) to show how the components A_r and A_z are projected upon the extension of R and a line tangent to the spherical surface through point P to be used for summing up to obtain the vertical components of acceleration A_V and the northward component A_N in the spherical coordinate system.

and

$$A_V = A_z \sin\vartheta + A_r \cos\vartheta.$$

Carrying out these manipulations we find

$$A_N = (\ddot{R}\sin\vartheta + 2\dot{R}\cos\vartheta\,\dot\vartheta - R\sin\vartheta\,\dot\vartheta^2 + R\cos\vartheta\,\ddot\vartheta)\cos\vartheta$$

$$+ (-\ddot{R}\cos\vartheta + 2\dot{R}\sin\vartheta\,\dot\vartheta + R\cos\vartheta\,\dot\vartheta^2 + R\cos\vartheta\,\dot\phi^2)\sin\vartheta$$

$$= 2\dot{R}\dot\vartheta + R\ddot\vartheta + R\sin\vartheta\cos\vartheta\,\dot\phi^2, \tag{6.9}$$

and

$$A_V = (\ddot{R}\sin\vartheta + 2\dot{R}\cos\vartheta\,\dot\vartheta - R\sin\vartheta\,\dot\vartheta^2 + R\cos\vartheta\,\ddot\vartheta)\sin\vartheta$$

$$+ (\ddot{R}\cos\vartheta - 2\dot{R}\sin\vartheta\,\dot\vartheta - R\cos\vartheta\,\dot\vartheta^2$$

$$- R\sin\vartheta\,\ddot\vartheta - R\cos\vartheta\,\dot\phi^2)\cos\vartheta$$

$$= \ddot{R} - R\dot\vartheta^2 - R\cos^2\vartheta\,\dot\phi^2. \tag{6.10}$$

Now we have some quite complicated forms of expression for accelerations in this spherical system, but we shouldn't be surprised after our experience of the new terms that came in even on the plane by shifting from rectilinear to polar forms.

6.2 ALTERNATIVE FORMS IN INERTIAL SPACE

Gathering these three components of the acceleration in spherical coordinates together in conventional order we have

Eastward: $\quad R\cos\vartheta\,\ddot{\phi} + 2\dot{\phi}(\dot{R}\cos\vartheta - R\sin\vartheta\,\dot{\vartheta})$ (6.11)

Northward: $\quad R\ddot{\vartheta} + 2\dot{R}\dot{\vartheta} + R\sin\vartheta\cos\vartheta\,\dot{\phi}^2$ (6.12)

Upward: $\quad \ddot{R} - R\dot{\vartheta}^2 - R\cos^2\vartheta\,\dot{\phi}^2.$ (6.13)

An alternative form can be written more concisely. If we define the velocities $u = R\cos\vartheta\,\dot{\phi}$, $v = R\dot{\vartheta}$, $w = \dot{R}$, then we have:

Eastward: $\qquad \dot{u} - \dfrac{uv}{R}\tan\vartheta + \dfrac{uw}{R}$ (6.14)

Northward: $\qquad \dot{v} + \dfrac{u^2}{R}\tan\vartheta + \dfrac{vw}{R}$ (6.15)

Upward: $\qquad \dot{w} - \dfrac{u^2 + v^2}{R}.$ (6.16)

If you want to give those complicated terms in the accelerations other than the doubly dotted ones some physical interpretation, like centrifugal forces, etc., you can work backward through the previous construction.

6.3 ACCELERATION AND CORIOLIS FORCES IN ROTATING SPHERICAL COORDINATES

We can obtain the expressions for the acceleration in coordinates rotating with uniform angular velocity Ω by substituting $\phi = \phi' + \Omega t$ into expressions (6.11–6.13). We find

$$R\cos\vartheta\,\ddot{\phi}' + 2(\dot{\phi}' + \Omega)(\dot{R}\cos\vartheta - R\sin\vartheta\,\dot{\vartheta}) = F_E', \quad (6.17)$$

$$R\ddot{\vartheta} + 2\dot{R}\dot{\vartheta} + R(\dot{\phi}' + \Omega)^2\sin\vartheta\cos\vartheta = F_N', \quad (6.18)$$

and

$$\ddot{R} - R\dot{\vartheta}^2 - R(\dot{\phi}' + \Omega)^2\cos^2\vartheta = -g + F_V', \qquad (6.19)$$

where we have written real force components F_E' (eastward), F_N' (northward), F_V' (vertical), [these forces are defined in the rotating system] and gravity g on the right hand side, as a separate force.

Writing the left hand side in conventional form

$$R\cos\vartheta\ddot{\phi} + 2\dot{\phi}'\dot{R}\cos\vartheta - 2\dot{\phi}'R\sin\vartheta\dot{\vartheta}$$
$$= -2\Omega\cos\vartheta\dot{R} + 2\Omega\sin\vartheta R\dot{\vartheta} + F_E' \quad (6.22)$$

$$R\ddot{\vartheta} + 2\dot{R}\dot{\vartheta} + R\sin\vartheta\cos\vartheta\dot{\phi}'^2$$
$$= -2\Omega\sin\vartheta R\cos\vartheta\dot{\phi} - \Omega^2 R\sin\vartheta\cos\vartheta + F_N' \quad (6.23)$$

$$\ddot{R} - R\dot{\vartheta}^2 - R\cos^2\vartheta\dot{\phi}'^2$$
$$= 2\Omega R\cos^2\vartheta\dot{\phi}' - g + \Omega^2 R\cos^2\vartheta + F_V'. \quad (6.24)$$

We find two kinds of virtual force on the right hand side. In order to make matters clearer we make use of the following substitutions

$$u' = R\cos\vartheta\dot{\phi}', \qquad v' = R\dot{\vartheta}, \qquad w' = \dot{R}$$

to obtain

$$\dot{u}' - \frac{u'v'}{R}\tan\vartheta + \frac{u'w'}{R}$$
$$= -2\Omega\cos\vartheta w' + 2\Omega\sin\vartheta v' + F_E', \quad (6.25)$$

$$\dot{v}' + \frac{u'^2}{R}\tan\vartheta + \frac{v'w'}{R}$$
$$= -2\Omega\sin\vartheta u' - \Omega^2 R\sin\vartheta\cos\vartheta + F_N', \quad (6.26)$$

and

$$\dot{w}' - \frac{u'^2 + v'^2}{R} = 2\Omega\cos\vartheta u' + \Omega^2 R\cos^2\vartheta + F_v' - g. \quad (6.27)$$

The terms with the factor 2Ω are all components of Coriolis force. The eastward momentum equation contains an additional Coriolis force that depends upon the vertical component of velocity w'. It has a Coriolis parameter that is different from those involving the velocity components u' and v'. It is $2\Omega \cos \vartheta$ instead of $2\Omega \sin \vartheta$. The vertical momentum equation also contains a new Coriolis force with this same new parameter multiplied by the eastward component of velocity.

The term $- R \sin \vartheta \cos \vartheta \, \Omega^2$ is an equatorward component of the centrifugal force of the rotating system. For equilibrium (and the happiness of a population of experts) this would have to be balanced by a real poleward force.

The term in the vertical equation, $R \cos^2\vartheta \, \Omega^2$, is a radially outward (vertical) component of the centrifugal force of the rotating system, but this can be absorbed easily into the gravitational force g by defining a new quantity called "gravity" g': $g' = g - R \cos^2\vartheta \, \Omega^2$.

It might not seem that such a complicated set of expressions could be useful, but we will soon see that they are.

6.4 TRAJECTORIES ON THE SURFACE OF A GRAVITATING SPHERE

In most dynamical problems these three expressions are written on the left-hand side of the equations of motion and the forces are written on the right-hand side. If the force acting on the particle is a gravitational force toward the center of a material sphere of constant radius a with a slippery frictionless surface and the surface of the sphere exerts a reaction F_R normal to the surface of the sphere (that is, radially), then the equations are simple. There are only two degrees of freedom (the particle moves along the surface with $R = a$).

$$\cos\vartheta\,\ddot{\phi} - 2\dot{\phi}\sin\vartheta\,\dot{\vartheta} = 0, \qquad (6.28)$$

$$\ddot{\vartheta} + \sin\vartheta\cos\vartheta\,\dot{\phi}^2 = 0, \qquad (6.29)$$

and

$$-\dot{\vartheta}^2 - \cos^2\vartheta\,\dot{\phi}^2 = -\frac{g}{a} + \frac{F_V}{a}. \qquad (6.30)$$

The third equation allows us to determine the reactive force F_V, which must be positive otherwise the particle leaves the sphere and executes an orbit in space. As long as the particle rides upon the sphere it does so in great circles. Since the sphere is slippery it doesn't really matter whether it rotates upon its axis or not. But the equations in this chapter are written in the absolute reference frame at rest with respect to the fixed stars. If we were to view the particle as it executes a great circle in absolute space from a reference frame rotating about the polar axis, these great circles would appear to execute pretty loops across the equator, as illustrated in some of the computer exercises. But these are not to be confused with motions referred to a cloud of expert particles rotating solidly. With great circles it is not possible to construct a cloud of expert particles that maintains its shape. One needs the possibility of steady orbits around the polar axis in small circles of latitude. Some mechanical contrivance is necessary for this type of solidly rotating configuration, much as we had to provide Hooke springs on the plane, or the combination of the sloping surface and vertical gravity in the parabolic dish.

As we will see in the next chapter a self-gravitating uniformly rotating mass of fluid particles can provide naturally a bulge at the equator that cancels the centrifugal force due to the mean rotation of the body, leaving only Coriolis forces that arise from velocities relative to the rotating body of other "expert" particles.

6.5 PLANAR MOTION IN SPHERICAL COORDINATES

We have just asserted that on a centrally gravitating perfect sphere, in absolute space, a sliding particle will always execute great circles. It seems physically obvious, but we want to offer here a more general and formal proof. If a particle moves under the action of a force directed only along the radius vector R so that $F_v \neq 0$ but $F_\vartheta = 0$ and $F_\phi = 0$, we expect the motion always to remain in the same plane, the one containing the initial position and velocity vectors. The exception is if the initial velocity is only radial, in which case the motion is confined to a line, not a plane, the line being the radius vector. This latter special case, of course, is excluded if the particle is confined to fixed radius. Now we wish to show from the statement $F_\vartheta = 0$ and $F_\phi = 0$ that we can actually deduce that the motion is in a fixed plane. For a particle constrained to move on a spherical surface this will lead to the fact that its motion is on a great circle, but the result is more general, for any case in free space, as long as the force is always toward the center. Note that if the particle is constrained to move on a nonspherical surface, the reactive force is necessarily noncentral, so that F_ϑ and F_ϕ won't generally vanish in that case, even if the external force is central.

We proceed as follows. First setting $F_\phi = 0$ in equation (6.21) gives us

$$R\ddot{\phi}\cos\vartheta + 2\dot{R}\dot{\phi}\cos\vartheta - 2R\dot{\phi}\dot{\vartheta}\sin\vartheta = 0. \quad (6.31)$$

Dividing by $R\dot{\phi}\cos\vartheta$ gives

$$\frac{\ddot{\phi}}{\dot{\phi}} + 2\frac{\dot{R}}{R} - 2\frac{\sin\vartheta\,\dot{\vartheta}}{\cos\vartheta} = 0, \quad (6.32)$$

which we recognize as

$$\frac{d}{dt}\ln(\dot{\phi}R^2\cos^2\vartheta) = 0, \quad (6.33)$$

or

$$\dot{\phi}R^2\cos^2\vartheta = \mu, \quad (6.34)$$

and this is our old friend $r^2\dot{\phi} = \mu$ where $r = R\cos\vartheta$ is the distance of the particle to the z axis; that is, we have conservation of angular momentum about the z axis. Setting $F_\vartheta = 0$ gives us

$$R\ddot{\vartheta} + 2\dot{R}\dot{\vartheta} + R\sin\vartheta\cos\vartheta\,\dot{\phi}^2 = 0. \quad (6.35)$$

Multiply by R and write this as

$$\frac{d}{dt}(R^2\dot{\vartheta}) + R^2\sin\vartheta\cos\vartheta\,\dot{\phi}^2 = 0. \quad (6.36)$$

Replace R^2 by $\mu/(\cos^2\vartheta\,\dot{\phi})$ to obtain

$$\frac{d}{dt}\left[\frac{1}{\cos^2\vartheta}\frac{\dot\vartheta}{\dot\phi}\right] + \tan\vartheta\,\dot\phi = 0. \qquad (6.37)$$

Finally, we can remove the explicit time dependence if we make the azimuth ϕ the independent variable. Then we have

$$\frac{d}{d\phi} = \frac{1}{\dot\phi}\frac{d}{dt}.$$

With the substitution (6.36) becomes the equation for the trajectory

$$\frac{d}{d\phi}\left[\sec^2\vartheta\frac{d\vartheta}{d\phi}\right] + \tan\vartheta = 0. \qquad (6.38)$$

But $\sec^2\vartheta$ is the derivative of $\tan\vartheta$, so we have

$$\frac{d^2}{d\phi^2}\tan\vartheta + \tan\vartheta = 0. \qquad (6.39)$$

This means $\tan\vartheta = a\cos\phi + b\sin\phi$ or back to rectilinear coordinates, $z = ax + by$ which is the equation for a plane through the origin. The reader may want to study and discuss the exceptional case in this derivation where we had a zero denominator.

PROBLEMS

Problem 6.0. A particle is found at 45° N latitude, and at rest relative to a rotating perfect sphere (not oblate). Calculate its time of crossing the equator following release.

EXERCISES

Exercise 6-1 [lines 34000–34270] *Drawing meridians and latitude circles on a spherical earth.* Many figures and illustrations of the stereographic projection are badly drawn even in scientific works of repute. Draftsmen seem to have strange ideas about how the ellipses that represent the meridians actually are supposed to look, and the pole is often incorrectly placed. You may want to make some drawings of phenomena on a stereographic projection of the sphere. This program will give you enough to work from. The inclination of the sphere can be set by the input [34020], which asks for the latitude and longitude that are to appear in the center of the projection. You can change the spacing of latitude circles by changing the STEP number in line 34050. You can change the spacing of the meridians by changing the STEP number in line 34110. Default values are set at 30 degrees. The STEP numbers in lines 34060 and 34100 set the density of points along the curves. The radius of the projection *R* is set at 99 in line 34005, you can change this too.

```
34000 ' Exercise 6-1. Draws spherical
34001 ' stereographic projection
34005 TPI=6.28318  :PI=TPI/2   :R= 99
34010 SCREEN 1: COLOR 0,2: KEY OFF:CLS
34020 INPUT "center lat,long"; LAT0, LON0
34021 LAT0= LAT0*TPI/360:LON0=LON0*TPI/360
34030 CLS
34040  INC = LAT0  : GOSUB 34160
34050 FOR I  = -90 TO 90  STEP 30
34060 FOR J  = 0 TO 360 STEP 2
34070 LAT = I: LON = J
34080 GOSUB 34220
34090 NEXT J  : NEXT I
34100 FOR I  = -90 TO 90  STEP 1
34110 FOR J  = 0 TO 360 STEP 30
34120 LAT = I: LON = J
34130 GOSUB 34220
34140 NEXT J  : NEXT I
34150 LOCATE 1,35:PRINT "end ": GOTO 34150
34160 FOR I = 1 TO 360 STEP 1
34170 A = I*TPI/360
34180 PSET (150+R*COS(A),100-R*SIN(A)),1
34190 NEXT
34200 LOCATE 1,35: PRINT "WAIT"
34210 RETURN
34220 LAT = LAT*PI/180:LON=LON*PI/180
34230 RHO=COS(LAT)
34240 X=RHO*COS(LON-LON0)
34241 Y=RHO*SIN(LON-LON0)*SIN(INC)
34250 Z=Y+SIN(LAT)*COS(INC)
34260 PSET(150+R*X,100-R*Z),1
34270 RETURN
```

Exercise 6-2 [lines 33000–33380] *A High Tower of Pisa.* This is a tower exercise only in the sense that a particle is released from a point exterior to a spherical earth, at a radius R that exceeds the earth's radius A, and allowed to fall freely under a central Newtonian gravitation varying inversely as the square of the distance from the center. This is a considerable simplification of the problem for the real earth, which is neither spherical nor with centrally directed gravitation (see chapter 9 for the Maclaurin ellipsoid). The initial latitude $TH = \vartheta$ and longitude $PH = \phi$ are given. Also the initial velocities are $THD = \dot{\vartheta} = 0$, $RD = \dot{R} = 0$, and $PHD = \dot{\phi} = \Omega = $ OMEGA. The OMEGA is the angular velocity of the rotating earth, which the particle at the top of a real tower would possess before release. The equations of motion, with vertical gravitational acceleration $G*A\hat{\ }2/R\hat{\ }2$, inserted in the equation for the vertical acceleration are equations (6.11–13). These are numerically integrated until $R < A$, at which particle has struck the earth. The lapsed time T, latitude, and longitude (corrected for the rotation of the earth—OMEGA*T) are then printed out, to show when and where the particle strikes the earth.

Appropriate parameters for gravity G, earth's radius A, and rotation rate OMEGA, etc. are specified in line 33020. A window is fixed [33025] to separate the geographical grid from the text. The rectangular geographical map grid, with 30 degree spacing in both latitude and longitude, is drawn in lines 33030–33050. An input for choice of latitude and ratio of initial radius to earth radius R/A, and conversion of the angles to radians are made in line 33060. The initial value of the longitudinal angular velocity PHD is fixed [33070].

The numerical integration of the equations of motion [33080–33130] is followed by definition of plotting variables L and M and plotting of the particle's geographical coordinates [33140–33160], and some printing on the screen of variables [33170–33210]. A graded time step is provided [33220–33240] as the particle approaches the earth's surface, at which point [33250] the integration process stops and a general printout [33270–33380] occurs. Before the impact, the integration is repeated cyclically by the instruction in line 33260, which repeats the integration steps. This program actually computes the Keplerian orbit of the dropping particle, but displays the result in the rotating reference geographical frame. The power of the computation considerably extends the range of study of the classical tower of Pisa problem [for a 60 meter tower, $R/A = 1.00001$—and be sure to make more places available in the printouts for PH—$OMEGA*T$ in lines 33180 and 33330, and reduce $DT0$ from the default value $DT0 = 60$ seconds (33020] to something like $dt0 = .02$. See sections 7.4 to 7.6 for an analytical approach and a discussion of the location of the base of the tower. The default values are chosen for really tall towers like $R/A = 3$ to 7. For example you can show that with input LAT0 = 60 (North), $R/A = 5.2$, the particle hits the earth at latitude 21 North, 31 degrees of longitude further east, after flight time of 3 hours and 4 minutes. This corresponds to releasing a mail bag, for example, high over Iceland, and having it land in Libya. Or, releasing it at $R/A = 6.7$ over the South Shetlands will cause it, after five hours' flight to swoop in low from the west, to land at Moscow. Or you can study orbits close to the geostationary one near lat = 0, $R/A = 6.2$.

If you prefer to observe the tracks on a nonrotating geographical grid, but with the tower's top still rotating at rate OMEGA, then you can change the definition of L in line 33140 to $L = (180/PI)*PH$. The orbits will now appear to be much more simple to the eye.

```
33000 'Exercise 6-2.  High tower of pisa
33010 SCREEN 1:COLOR 0,2:KEY OFF
33011 CLS :PI=4*ATN(1)
33020 G = 9.8: A =6378000!
33021 OMEGA =2*PI/86000!:DT0=60 :DT=DT0
33030 FOR I = 0 TO 360 STEP 30
33031 FOR J=-90 TO 90 STEP 30
33035 WINDOW SCREEN (0,-50)-(400,200)
33040 LINE(0,100)-(I,100-J),1,B
33050 NEXT:NEXT
33060 INPUT "start lat,altitude R/a";LAT0,R0
33061 R=A*R0:TH =LAT0*PI/180
33070 PHD=OMEGA
33080 AAA=R*(COS(TH))^2*PHD^2-G*A^2/(R^2)
33081 RDD=R*THD^2+AAA
33090 RD=RDD*DT+RD: R=RD*DT+R
33100 AAA=(-2*RD*THD-R*SIN(TH)*COS(TH)*PHD^2)
33101 THDD= (1/R)*AAA
33110 THD=THDD*DT+THD: TH= THD*DT+TH
33120 AAA=-2*PHD*(RD*COS(TH)-R*SIN(TH)*THD)
33121 PHDD= (1/(R*COS(TH)))*AAA
33130 PHD=PHDD*DT+PHD:PH=PHD*DT+PH
33140 L =( 180/PI)*(PH-OMEGA*T)
33141 M=100-(180/PI)*TH
33150 T = T+DT
33160 PSET(L,M),7
33170 LOCATE 2,1
33180 AA=TH*180/PI:BB=( PH-OMEGA*T)*180/PI
33181 PRINT USING "lat=##.#,lon=###.#";AA,BB
33190 LOCATE 3,1
33191 PRINT USING "r/a=##.####";R/A
33192 LOCATE 3,13
33193 PRINT USING " t= ###### seconds"; T
33200 IHRS= INT(T/3600):DHRS=(T/3600)-IHRS
33201 IMIN=INT(DHRS*60)
```

```
33220 IF R/A<1.5 THEN DT = .3*DT0
33230 IF R/A<1.1  THEN DT = .1*DT0
33240 IF R/A<1.02 THEN DT = .01*DT0
33250 IF R/A<1 THEN BEEP:GOTO 33270
33260 GOTO 33080
33270 LPRINT " "
33280 LPRINT "Exercise 6-2"
33290 LPRINT "omega ";OMEGA
33300 LPRINT "initial values:"
33310 LPRINT USING "lat ##,r/a ##.###";LAT0,R0
33320 LPRINT "final values:"
33330 AA=TH*180/PI:BB=(PH-OMEGA*T)*180/PI
33331 LPRINT USING "lat=##.# lon=###.#";AA,BB
33340 LPRINT USING "r/a=##.###";R/A
33341 LPRINT USING "t=#### sec";T
33350 LPRINT USING "t ###h ##m";IHRS,IMIN
33360 LPRINT " "
33370 LPRINT " "
33380 END
```

REFERENCES

Batchelor, G. K. 1967. *An Introduction to Fluid Dynamics.* Cambridge: Cambridge University Press (615 pp).

Morse, P. M. and H. Feshbach. 1957. *Methods in Mathematical Physics.* New York: McGraw Hill (2 vols., 1978 pp.).

CHAPTER VII

Huygens' rotating oblate earth

7.1 APPROXIMATE FIGURE OF THE EARTH

In this chapter we introduce a method of constructing a spheroidal platform, which will support a population of experts (sliding upon it in solid rotation at the same angular velocity) held to the surface by a centrally directed gravitational force, computed as though all the mass of the earth were concentrated at its center. The platform itself, though doubtless a major engineering structure, encompassing as it does the whole earth's surface, is viewed as massless, as far as the computation of the gravitational attraction is concerned. As we will see, this is a bad approximation for computing the figure of the earth and leads to a theoretical equatorial bulge only about half that of the earth. Nevertheless, the equations of motion for the particle that are derived are very close to the ones derived (approximately) for the more physically realistic construction (chapter 9), in which the mass of the bulge is permitted to have an effect on the direction and strength of gravitational attraction. The subject of computing the figure of the earth from first principles has been elaborated extensively over the past three centuries, starting with Newton, Clairaut, and Laplace and exercising the interest and ingenuity of more recent renowned theorists such as

Kelvin, Poincaré, and Chandrasekhar. In an elementary introduction we are going to have to find ways of skirting the difficult sophisticated aspects of the problem of rotating gravitating figures of equilibrium, but still reaching the goal of getting reliable dynamical equations for motion of a particle on the earth. The simple Huygens spheroid shows the way.

7.2 FORCE ON A PLUMB BOB

To be sure that we have a clear picture of the forces acting on a particle someplace on the rotating earth, let us consider a plumb bob B (figure 7.1) of unit mass located at the earth's radius $R = a$ and geocentric latitude ϑ. The distance from the axis of rotation is $r = a \cos \vartheta$. The bob is suspended by a short string PB from the point P elevated above the earth's surface. Extend the line PB to the equatorial plane where it intersects at point A. The angle $\vartheta + \delta$ is the latitude that would be measured astronomically, and the angle δ is the difference between the two kinds of latitude mentioned. It may be thought of as a small deviation of the local direction of the vertical at B as given by the plumb line PB from the direction BO toward the center of the earth's attraction. We remember that a true spheroid's attraction is only approximately toward the center, but we neglect this here, in the spirit of Huygens. This construction is shown in panel a.

Inasmuch as the bob is rotating around the earth's axis CO at a distance BC, it is in relative equilibrium and is accelerating toward C by the amount $\Omega^2 r = \Omega^2 a \cos \vartheta$. This acceleration is produced by the resultant of the only two forces acting upon the bob: central gravitation g and T, the tension in the string. The observer on the rotating

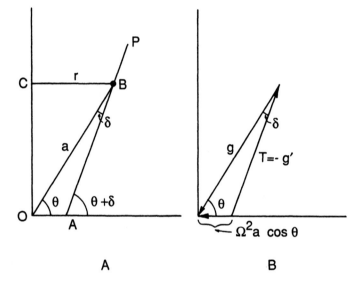

Figure 7.1. Construction to show balance of forces on a pendulum bob *B* a distance *a* from the center of the earth *O* and at latitude ϑ. On the left panel (a) the bob is shown rotating with the earth around a point *C* on the earth's axis, *CO*. The bob is suspended from point *P*, where *PB* is very short compared to the other scale lengths of the problem. On the right panel (b) the triangle of forces is drawn. The $\Omega^2 a \cos \vartheta$ force is virtual.

earth will interpret this tension as a measure of local gravity (here the word gravity is used instead of gravitation), which we can denote by g'. The force diagram is shown in panel b. To compute the deviation δ we make use of the law of sines:

$$\sin\delta/\sin(\vartheta + \delta) = \Omega^2 a\cos\vartheta/g. \qquad (7.1)$$

If we take the following terrestrial values for the constants in the problem $2\pi/\Omega = 8.61 \times 10^4$ sec, $a = 6.38 \times 10^6$ m, $g = 9.8$ m s^{-2} we find $\Omega^2 a/g + 0.0035$, a small quantity. We obtain $\delta \cong \sin\delta \cong (\Omega^2 a/g)\sin\vartheta\cos\vartheta$. The maximum is at 45° latitude, where δ is only about 6 minutes of arc.

7.3 COMPUTING THE BULGE

The surface of the platform that we construct on the earth is normal to the distribution of apparent gravity, so the radius R can be determined from

$$\frac{dR}{Rd\vartheta} = -\tan\delta \cong -\delta. \qquad (7.2)$$

Thus

$$\frac{dR}{d\vartheta} = -\frac{\Omega^2 a^2}{g}\sin\vartheta\cos\vartheta, \qquad (7.3)$$

and

$$R = a + \frac{\Omega^2 a^2}{2g}\cos^2\vartheta. \qquad (7.4)$$

The radius R of our platform at the equator thus exceeds that at the pole by $\Omega^2 a^2/2g$, which using our constants is 11 km, an unrealistically small value.

The tension $T = -g'$ can also be used by the law of sines to obtain

$$g/g' = \sin(\vartheta + \delta)/\sin\vartheta$$

$$= \frac{\Omega^2 a}{g}\cos^2\vartheta + \cos\delta, \qquad (7.5)$$

or

$$g' \cong g\left(1 - \frac{\Omega^2 a}{g}\cos^2\vartheta\right), \qquad (7.6)$$

so that this enables us to make a first estimate of the variability of apparent local gravity with latitude over the earth.

Here is another way to compute the bulge. We assume that in absolute space there is an expert particle with $\dot{\vartheta}_e = 0$ and $\dot{\phi}_e = \Omega$, where the angular velocity Ω is a constant independent of ϑ and ϕ. The equation (6.17) is therefore satisfied identically, and the two equations (6.18–6.19) then are:

$$\Omega^2 \sin\vartheta\cos\vartheta = F_N/a \qquad (7.7)$$

and

$$\Omega^2\cos^2\vartheta = (g - F_V)/a. \qquad (7.8)$$

The force components F_N (added to 6.18) and F_V are the forces necessary for $\dot{\vartheta} = 0$ and $\dot{\phi} = \Omega$. In the absence of the force F_N the particles would oscillate across the equator.

The most convenient way to obtain the force F_N is to build an equatorial bulge on the sphere so that the radius is not a true constant but has the shape $R = a + b\cos^2\vartheta$. The poleward slope of this surface is then $-(2b/a)\sin\vartheta\cos\vartheta$ and this provides a poleward component of gravity of $g(2b/a)\sin\vartheta\cos\vartheta$.

If we choose the amplitude of the bulge as $b = a^2\Omega^2/2g$ we have now got a northward force F_N that can permit a cloud of experts to rotate solidly about the axis at angular velocity Ω. The last equation is really just for determining the reactive force F_v, which means that the extra term $-a\,\Omega^2\cos^2\vartheta$ can be absorbed into a redefinition of gravity g', which now, in contradistinction to gravitation g, is weaker at the equator than at the poles [equation (7.5)]. This oblate spheroid is a reference frame that can support a population of expert particles sliding around in latitude circles around the polar axis, all in solid rotation with angular velocity Ω.

7.4 NOVICE PARTICLES ON HUYGEN'S SPHEROID

The equations for a novice particle in the inertial frame then are

$$\cos\vartheta\,\ddot\phi - 2\dot\phi\sin\vartheta\,\dot\vartheta = 0,$$
$$\ddot\vartheta + \dot\phi^2\sin\vartheta\cos\vartheta = \Omega^2\sin\vartheta\cos\vartheta,$$

and

$$\dot\vartheta^2 + \cos^2\vartheta\,\dot\phi^2 = \Omega^2\cos^2\vartheta + (g - F_v)/a. \qquad (7.9)$$

The vertical accelerations have been neglected. The last equation can be used to calculate the reactive force.

We now come to the concept of a Coriolis force again. Writing the equations (7.9) with reference to the cloud of experts requires only the replacement of the variable for absolute longitude ϕ with the longitude referred to a fixed longitude on earth [Greenwich]: $\phi' = \phi - \Omega t$.

$$\cos\vartheta\,\ddot{\phi}' - 2(\dot{\phi}' + \Omega)\sin\vartheta\,\dot{\vartheta} = 0$$

and

$$\ddot{\vartheta} + (\dot{\phi}' + \Omega)^2\cos\vartheta\sin\vartheta = \Omega^2\sin\vartheta\cos\vartheta. \quad (7.10)$$

Cancelling out the terms that balance for the absolute reference frame in the second equation becomes

$$\ddot{\vartheta} + \sin\vartheta\cos\vartheta\,\dot{\phi}'^2 + 2\Omega\dot{\phi}'\sin\vartheta\cos\vartheta = 0. \quad (7.11)$$

Now we can place all terms that do not correspond to the standard form of the expression for acceleration in spherical coordinates on the right hand side, and multiply both equations by the earth's mean radius a:

$$a\cos\vartheta\,\ddot{\phi}' - 2\dot{\phi}'a\sin\vartheta\,\dot{\vartheta} = 2\Omega a\sin\vartheta\,\dot{\vartheta}$$

and

$$a\ddot{\vartheta} + a\sin\vartheta\cos\vartheta\,\dot{\phi}'^2 = -2\Omega a\sin\vartheta\cos\vartheta\,\dot{\phi}'. \quad (7.12)$$

Defining relative velocities $u' = a\cos\vartheta\,\dot{\phi}'$ and $v' = a\dot{\vartheta}$ the two terms on the right hand side can be written as $2\Omega\sin\vartheta\,v'$ and $-2\Omega\sin\vartheta\,u'$. Thus, in the relative frame of the rotating earth Coriolis forces make their appearance again, in a different form from that in the Hooke plane. The Coriolis force acts to the right (in Northern hemisphere) of the direction of the horizontal components of velocity. The Coriolis parameter is not 2Ω but $2\Omega \sin \vartheta$, and is a function of latitude. In general the relative novice trajectories computed from (7.12) on the spheroid are not circles. If the amplitude of the relative motion is small, then neglecting quadratic terms $\dot{\phi}'\dot{\vartheta}'$ and $\dot{\phi}'^2$ we can write

$$\dot{u}' = 2\Omega \sin \vartheta\, v',$$

and

$$\dot{v}' = -2\Omega \sin \vartheta\, u', \qquad (7.13)$$

so that the relative orbits are circles with frequency $2\Omega \sin \vartheta$. This is the same dependence of frequency on latitude that we found for small amplitude relative motions on the paraboloid of section 4.2.

If the relative velocity of the novice particle is sufficiently great its trajectory will cover a significant range of latitude and the Coriolis parameter will vary along the path of the particle. The trajectory will then no longer be a circle. The types of path are explored in the computer exercises.

For most terrestrial types of problem a particle on the spheroidal platform will never slide fast enough to reduce the reactive force F_r to zero, at which time it would have to lift off the surface into free space. With centrally directed gravity it would then execute a Keplerian ellipse, whose elements would depend upon the absolute velocity at liftoff. If one wants to think about particles originally at rest in the earth's coordinate system, but not initially in contact with the surface (for example, a particle dropped from a tower), approximate solutions can be obtained using the spherical coordinates expressions for acceleration in chapter 6 (equations 6.11–13), introducing a centrally directed gravitational force in the equation for the vertical acceleration, and subtracting from ϕ the amount of the earth's rotation Ωt, thus obtaining the equations for the rotating frame. Problems involving orbits of satellites launched from the earth are really more complicated than this because of the non-central nature of the earth's real gravitational field—the bulge actually affects the orbits, and they are not true Keplerian ellipses.

The reader will scarcely fail to notice that in the development of our ideas of a platform so far we have been at some pains to distinguish between the case where the particle is moving in two dimensions along a plane or conceptually rigid platform and the case where the particle departs from it and moves in free space. In the latter case the platform exerts no reactive force to counterbalance the centrifugal force due to the rotation of the rotating coordinate frame, and so one may wonder at computing its motion in the moving frame. However, when we think of the ocean or atmosphere particles are not confined to a rigid platform, and in fact can move, within limits, in three dimensions. In chapter 9 we will make a brief foray into this more general state of affairs, when we have defined the idea of equipotential surfaces and hydrostatic equilibrium of a rotating fluid mass. We will then discover that we have a more general three dimensional equilibrium that cancels out the centrifugal force due to the rotation of the coordinate system, and introduces another Coriolis force associated with vertical velocity of a particle. Of course, we already have a foretaste of this in the general form of equations for the accelerations in spherical coordinates given in chapter 6, but as yet not implemented in any concrete example.

7.5 FREE FALL FROM A SHORT TOWER

We now consider a famous problem of determining where a particle, dropped from the top of a tower, will strike the surface of the earth. It is often spoken of as the Tower of Pisa problem, presumably in honor of Galileo, who is said to have conducted some fundamental experiments there that helped establish the laws of dynamics. To the approximation that interested Galileo the effects of the earth's rotation were negligible. We want to deduce how much the earth's rotation deflects the trajectory of the falling particle from the local vertical, as determined from a plumb bob suspended from the point of release at the top of the tower. It isn't very much for the 50 meter high Tower of Pisa, but determining it is a good exercise in clarifying our ideas about dynamics on a rotating spheroid. First we note that for a nonrotating spherical earth, with gravitational attraction toward the center, the particle will fall directly toward the center, and strike the earth's surface precisely at a point indicated by a plumb bob suspended from the point of release. The plumb bob, of course, defines the local vertical. The time t_1 for the fall is calculated from the vertical equation of motion,

$$\ddot{z} = -g$$

where z is the vertical distance of the particle above the earth's surface, and g is the local value of gravity taken to be a constant. Integrating twice

$$\dot{z} = -gt, \qquad z = h - \tfrac{1}{2}gt^2$$

where at $t = 0$ we assume $z = h$ and $\dot{z} = 0$. From the last expression we find the time of impact $t_1 = \sqrt{2h/g}$.

If the earth is rotating, the particle at the top of the tower actually has an eastward velocity u in inertial space that amounts to $\Omega R_0 \cos \vartheta$ where R_0 is the local radius of the spheroid and ϑ is the geocentric latitude. While falling the particle is *unsupported* by any reactive force from contact with the earth's surface such as might be influenced by the equatorial bulge of the Huygens spheroid. Therefore one might expect that it would be flung outward from the earth's axis by the centrifugal force of the system while falling, and that its point of impact with the surface would be deflected equatorward. Also, from the conservation of angular momentum, we notice that the base of the tower is closer (by an amount $h \cos \vartheta$) to the earth's axis than the top of the tower, so the particle might speed up and outrun the eastward movement of the base of the tower, thus striking the earth somewhat to the east of the tower. So our preliminary physical analysis suggests that the particle is deflected both southward and eastward from the base of the tower in the course of its descent. As we will see in the next two sections this is not entirely correct, the trouble being that we have to be careful about our definition of the location of the base of the tower. As determined by a plumb bob on a rotating earth, the base of the tower lies somewhat equatorward of the point intersected on the earth's surface by a radius drawn to the earth's center from the top of the tower.

We investigate the Tower of Pisa problem in two different ways, and we hope that by contrasting the two the reader will recognize how much easier it is to do the problem in the rotating reference frame than in the inertial reference frame. Indeed, the main reason for introducing rotating reference frames in problems is to simplify the dynamical analysis. Some problems of course are so very simple that they are most easily done in inertial space, but not all. When studying the atmosphere and ocean it is almost necessary to refer to rotating coordinates, because the forces, like pressure gradients, are naturally defined with reference to them.

7.6 CALCULATION OF THE DEFLECTION OF A FALLING PARTICLE IN A ROTATING COORDINATE FRAME.

Let us refer to the equations of motion (6.25–6.27) for a particle in spherical coordinates in a rotating frame. We can assume that with a short tower ($h \ll R$) the relative radial component of velocity of fall, w', greatly exceeds any relative horizontal components of velocity u', v' that are induced during the fall. With only gravity g' acting ($F_E' = F_N' = F_v' = 0$) the leading terms in each of these equations are

$$\dot{u}' = -2\Omega\cos\vartheta w', \qquad \text{(eastward)}$$

$$\dot{v}' = -R\sin\vartheta\cos\vartheta\,\Omega^2, \qquad \text{(northward)}$$

and

$$\dot{w}' = R\cos^2\vartheta\,\Omega^2 - g \equiv -g'. \qquad \text{(radially)}$$

Begin by calculating $z'(t)$ from the third equation, where z' is the radially outward coordinate ($z = 0$ at the earth's surface, $z = h$ at top of the tower). At $t = 0$, $x' = 0$, $y' = 0$, $z' = h$, $w' = 0$; then integrating

$$w' = -g't, \qquad z' = h - \tfrac{1}{2}g't^2,$$

and the time t_1 of impact is

$$t_1 = \sqrt{2h/g'}.$$

From the first equation we can now introduce the expression for w' to obtain

$$u' = -2\Omega\cos\vartheta(-g't^2/2),$$
$$x' = 2\Omega\cos\vartheta g't^3/6,$$

and

$$x'(t_1) = 2\Omega\cos\vartheta\frac{g'}{6}(\sqrt{2h/g'})^3,$$

where $x'(t_1)$ is the x' displacement on the sphere. From the second equation we have

$$v' = -R\sin\vartheta\cos\vartheta\,\Omega^2 t,$$
$$y' = -R\sin\vartheta\cos\vartheta\,\Omega^2 t^2/2,$$

and

$$y'(t_1) = -R\sin\vartheta\cos\vartheta\,\Omega^2 h/g',$$

where $y'(t_1)$ is the northward displacement (negative) in the system. Now we must not jump at the conclusion that the particle will strike the spherical surface to the east and south of the tower, because we haven't determined the position of the base of the tower yet. We do that by suspending a plumb bob from the top of the tower. Now we have introduced a reactive force due to the tension in the string, attached to the top of tower, and rotating in equilibrium with the earth. From equation (7.1) we find the angle δ,

$$\delta \cong \sin\delta \cong \Omega^2 R \sin\vartheta \cos\vartheta / g',$$

and therefore the base of the tower is located at $x_B' = 0$, and

$$y_B' = -\Omega^2 R \sin\vartheta \cos\vartheta\, h/g'.$$

Therefore we have immediately

$$y'(t_1) - y_B' = 0$$

and we have the result that there is no southward displacement of the point of impact from the base of the tower. At 45°N on the Huygens spheroid the angle δ is about 0.00175 radians, so with tower height of 50 meters, the two balancing southward displacements are not negligible. The eastward $x'(t_1)$ displacement is calculated to be 1.1 centimeters.

7.7 FALL FROM A TOWER CALCULATED IN INERTIAL SPACE.

7.7a Preliminary results regarding ellipses.

Let ξ, η be Cartesian coordinates of points in a plane, with origin at the center of the earth, C, and let r, ψ be polar coordinates in the same plane. Then

$$\xi = r\cos\psi, \qquad \eta = r\sin\psi.$$

We wish to describe an ellipse with focus at the origin, and major axis of length, $2a$, coincident with the ξ axis.

Let e be the eccentricity of the ellipse. Then the center of the ellipse is at

$$\xi = ae, \eta = 0$$

and the other focus is at $\xi = 2ae$, $\eta = 0$. See figure 7.2.

The sum of the distances to the two foci from any point P on the ellipse is equal to $2a$, by definition. If ξ, η are the coordinates of P then we have

$$\sqrt{\xi^2 + \eta^2} + \sqrt{(\xi - 2ae)^2 + \eta^2} = 2a.$$

A little algebra shows that this may be rewritten in the polar coordinates as

$$r = \frac{a(1 - e^2)}{1 - e\cos\psi}, \tag{7.7.1}$$

and in the Cartesian coordinates as

$$\frac{(\xi - ae)^2}{a^2} + \frac{\eta^2}{a^2(1 - e^2)} = 1. \tag{7.7.2}$$

From (7.7.2) we see that the ellipse can also be represented by $\xi - ae = a \cos \gamma$ and $\eta = a\sqrt{1 - e^2}\sin \gamma$. This representation may be visualized as follows. Construct a circle of radius a with center at $(ae, 0)$. See figure 7.2A. Let P' have coordinates (ξ, η') and lie on the circle. Then $\xi - ae = a \cos \gamma$ and $\eta' = a \sin \gamma$ is a parametric representation of this circle in terms of angle γ measured from the direction of the ξ axis. Now if we compress the circle in the η direction by a factor $\sqrt{1 - e^2}$, so that $\eta = \sqrt{1 - e^2}\eta'$, then the circle is mapped into our ellipse. See figure 7.2B.

Finally, we will need the area $A(\psi)$ swept out by the radius vector from the focus at C to P as ψ increases from

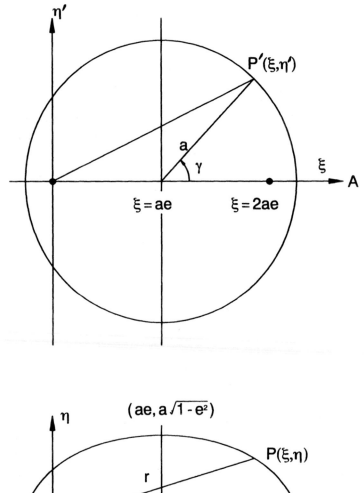

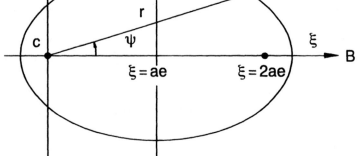

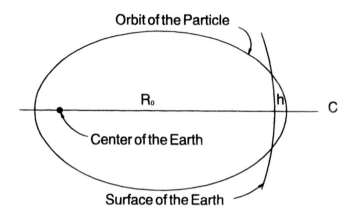

Figure 7.2. Panel B shows the ellipse of the particle's trajectory with focus at the origin $\xi = \eta = 0$ and illustrates the Cartesian (ξ, η) and polar (r, ψ) coordinates used to describe it. Panel A is the reference circle in (ξ, η') space used to compute the area $A(\psi)$. Panel C shows the configuration of the elliptical orbit with respect to the earth. The particle dropped from height h above the earth's surface hits the earth where the orbit intersects the earth's surface. In Panel C, the center of the earth is the origin $r = 0$.

0 to its present value. The result is easily found geometrically by considering the corresponding area in the circle, and then compressing it uniformly in η to get the result for the ellipse. We find

$$A(\psi) = \frac{1}{2}a^2\sqrt{1-e^2}[\gamma + e\sin\gamma], \qquad (7.7.3)$$

where

$$\sin\gamma = \frac{\sqrt{1-e^2}\ \sin\psi}{1 - e\cos\psi}. \qquad (7.7.4)$$

7.7b FREELY FALLING PARTICLE

We now consider the free fall of a particle in inertial space. Imagine that its position in spherical coordinates is initially at $R = R_0 + h$, the latitude $\vartheta = \vartheta_0$, and longitude $\phi = 0$. The initial angular velocity of the particle is $\dot{\phi} = \Omega$. The only force acting on the particle is gravitation, radially inward, of magnitude GM/R^2. Since the distance to the axis of rotation is $(R_0 + h)\cos\vartheta$, the initial velocity is $v_0 = (R_0 + h)\cos\vartheta\,\Omega$ toward the east (positive ϕ direction). In terms of a cartesian coordinate system with $z = R\sin\vartheta$, $y = R\cos\vartheta\sin\phi$, $x = R\cos\vartheta\cos\phi$ the plane containing the initial velocity vector and the origin is $z/x = \sin\vartheta_0/\cos\vartheta_0$ and the orbit of the particle always stays in this plane.

Let $\xi = x/\cos\vartheta_0 = z/\sin\vartheta_0$ and $\eta = y$ be cartesian coordinates in the plane of the orbit, and r, ψ be polar coordinates in the same plane. The initial conditions for the particle are $r = R_0 + h$ and $\dot{\psi} = \Omega\cos\vartheta$ since v_0 must be the same in both coordinate systems. Now we conserve energy and angular momentum for the particle, so

$$\frac{1}{2}(\dot{r}^2 + r^2\dot{\psi}^2) - \frac{GM}{r} = E,$$

and

$$r^2\dot{\psi} = \mu = (R_0 + h)^2\Omega\cos\vartheta_0.$$

The energy equation in terms of r alone is

$$\frac{1}{2}\left(\dot{r}^2 + \frac{\mu^2}{r^2}\right) - \frac{GM}{r} = E,$$

one time derivative of which gives

$$\ddot{r} - \frac{\mu^2}{r^3} + \frac{GM}{r^2} = 0. \tag{7.7.5}$$

Let $q = 1/r$ and use $r^2\dot{\psi} = \mu$ to write

$$\frac{d}{dt} = \mu q^2\frac{d}{d\psi}.$$

Substituting into the equation (7.7.5) we find

$$\mu q^2\frac{d}{d\psi}\left(\mu q^2\frac{d}{d\psi}\left(\frac{1}{q}\right)\right) - \mu^2 q^3 + GMq^2 = 0,$$

or simply

$$\frac{d^2 q}{d\psi^2} + q = \frac{GM}{\mu^2}. \tag{7.7.6}$$

At time $t = 0$ our initial conditions are $\psi = 0$, $r = R_0 + h$ and $\dot{r} = 0$. This means $q = 1/(R_0 + h)$ and $dq/d\psi = 0$ at $\psi = 0$. Therefore

$$q = \frac{GM}{\mu^2} + \left(\frac{1}{R_0 + h} - \frac{GM}{\mu^2}\right)\cos\psi$$

is the equation for the orbit. This means

$$r = \frac{\mu^2}{GM} \frac{1}{1 + \left(\dfrac{\mu^2}{GM(R_0 + h)} - 1\right)\cos\psi}. \qquad (7.7.7)$$

Comparison with equation (7.7.1) shows that this is an ellipse with eccentricity

$$e = 1 - \frac{\mu^2}{GM(R_0 + h)}. \qquad (7.7.8)$$

We write this as $e = 1 - \alpha$ where

$$\alpha = \frac{\mu^2}{GM(R_0 + h)} = \frac{(R_0 + h)^3 \Omega^2 \cos^2\vartheta_0}{GM}. \qquad (7.7.9)$$

The semi-major axis a is given by

$$a = \frac{\mu^2}{GM(1 - e^2)} = \frac{R_0 + h}{2 - \alpha}. \qquad (7.7.10)$$

Let the particle reach the surface $r = R_0$ at $\psi = \psi_1$. Now equation (7.7.7) may be written as

$$r(1 - e\cos\psi) = \alpha(R_0 + h),$$

where $\alpha = 1 + e$. Then we find that

$$\cos\psi_1 = 1 - \left(\frac{\alpha}{1 - \alpha}\right)\frac{h}{R_0}. \qquad (7.7.11)$$

From (7.7.11) we have

$$\sin\left(\frac{\psi_1}{2}\right) = \sqrt{\frac{h}{2R_0}\left(\frac{\alpha}{1-\alpha}\right)},\qquad(7.7.12)$$

and

$$\cos\left(\frac{\psi_1}{2}\right) = \sqrt{1 - \frac{h}{2R_0}\left(\frac{\alpha}{1-\alpha}\right)}.\qquad(7.7.13)$$

The angle γ_1 corresponding to ψ_1 may be found from equations (7.7.4). After some tedious algebra we find

$$\sin\gamma_1 = \frac{2}{(1 + h/R_0)}\sqrt{\frac{h}{R_0}\frac{(1-\alpha/2)}{(1-\alpha)}\left(1 - \frac{h}{2R_0}\left(\frac{\alpha}{1-\alpha}\right)\right)}.\qquad(7.7.14)$$

All of these results in equations (7.7.1) through (7.7.4) are exact, and do not involve assuming that either α or h/R_0 is small.

We need to go back to the original x, y, z or R, ϑ, ϕ system to find where the particle lands. In the original system $z = R \sin\vartheta$ and in the plane of the orbit $z = \xi \sin\vartheta_0 = r \sin\vartheta_0 \cos\psi$. So when $r = R = R_0$ the latitude ϑ_1 is given by

$$\sin\vartheta_1 = \sin\vartheta_0\cos\psi_1 = \sin\vartheta_0\left[1 - \frac{h}{R_0}\left(\frac{\alpha}{1-\alpha}\right)\right].\qquad(7.7.15)$$

This result is also exact. A little geometry for the short tower case ($h/R_0 \ll 1$) shows that the particle is displaced toward the equator by an amount

$$\frac{h\sin\vartheta_0}{\cos\vartheta_0}\left(\frac{\alpha}{1-\alpha}\right) = \frac{R_0\Omega^2\sin\vartheta_0\cos\vartheta_0}{g'}, \quad (7.7.16)$$

where $g' = (GM/R_0) - R_0\cos^2\vartheta_0\Omega = g(1-\alpha)$ is apparent gravity. This is the same result found in section 7.5.

More algebra shows that as long as γ_1 is small, the longitude of impact ϕ_1 is given by

$$\phi_1\cos\vartheta_0 = \sqrt{\frac{\alpha}{2-\alpha}}\sin\gamma_1$$

$$\times \left[1 + \frac{h}{R_0}\left(1 - \frac{\alpha}{3(1-\alpha)} - \frac{2}{3}\left(\frac{\alpha}{1-\alpha}\right)\tan^2\vartheta_0\right)\right]$$

$$(7.7.17)$$

where terms of order $(h/R_0)^2$ have been omitted. This is the longitude of impact in the inertial coordinate system. We really want $\phi_1' = \phi_1 - \Omega t$, which is what we see in the rotating system. For this we need the time of impact t_1.

The conservation of angular momentum in the plane of the orbit implies that the rate at which the area is swept out by the radius vector is a constant, equal to $\mu/2$. Therefore the particle lands at t_1 given by

$$\mu t_1 = a^2\sqrt{1-e^2}[\gamma_1 + e\sin\gamma_1]. \quad (7.7.18)$$

The need to compute this time is the real reason that we need the angle γ. More tedious algebra leads to

$$\Omega t_1\cos\vartheta_0 = \sqrt{\frac{\alpha}{2-\alpha}}\sin\gamma_1\left[1 + \frac{h}{R}\frac{1}{3(1-\alpha)}\right], \quad (7.7.19)$$

where again terms of the order $(h/R_0)^2$ are neglected.

Comparison of (7.7.17) and (7.7.19) shows that ϕ_1 and Ωt_1 are the same to leading order and that the difference is

$$[\phi_1 - \Omega t_1]\cos\vartheta_0 = \sqrt{\frac{\alpha}{2-\alpha}}\sin\gamma_1\left(\frac{2h}{3R_0}\right)$$

$$= \Omega t_1\left(\frac{2h}{3R_0}\right)\cos\vartheta_0, \qquad (7.7.20)$$

assuming α is small. Again the result agrees with the previous section.

The lesson that we learn from this calculation is obvious. The computations in this section are involved and tedious. The probability of making an error is large. The computations in section 7.6, using a coordinate system fixed relative to a rotating earth and including the idea of Coriolis forces are much simpler.

PROBLEMS

Problem 7.0. A particle is initially at rest on the rotating Hooke plane. It is attached to another point fixed in the rotating coordinate system of the plane by a Hooke spring that is stiffer than usual for the rest of the system. What is the qualitative nature of the orbit of the particle in the rotating reference system?

Problem 7.1. A truck is running south on slippery wintertime Route 385 (40°N latitude) at 20 meters per second. Calculate the Coriolis force (toward the west) per unit mass of truck, and by comparing it to gravity (9.8 m sec^{-2}) determine how much the highway would need to be banked from the vertical to counteract any tendency for the truck to drift off the road.

Problem 7.2. A particle is initially at rest in absolute space at 30°N latitude on the slippery surface of an oblate earth. Calculate its trajectory when released, and the time when it returns to its original point of rest.

Problem 7.3. In chapter 3 we posed a problem that involved introducing small perturbations to the Hooke spring dynamical system. Try the same thing with the particle sliding on our spheroidal platform.

Problem 7.4. Show that when a particle rotating in equilibrium around the earth's platform at some constant latitude has its energy perturbed by δE or its angular momentum perturbed by $\delta \mu$, its path wobbles around the sphere between two latitudes, and derive these latitudes from the energy equation.

Problem 7.5. In 1912, at a meeting of mathematicians in Cambridge, England, Dr. John Hagen of the Vatican Observatory reported on an experiment with an Atwood machine 20 meters high. The Atwood machine is a device for effectively reducing gravity in experiments simulating free fall. It consists of two equal weights that are suspended by a connecting line that hangs over a pulley. When a small extra weight is placed on one of them, the entire system begins to accelerate—the heavier weight going down, the lighter upward. Coriolis force can be deduced from the eastward deviation from the vertical of the string supporting the descending weight. Supposing that gravity has been reduced by ⅕, compute the expected horizontal displacement of the supporting string just before it arrives at the bottom of its 20 meter descent.

Problem 7.6. Write a program that will show you the trajectory of a particle initially sliding on a Hooke plane in an inertial circle, but subject to a resistive force proportional to the velocity relative to the rotating reference frame.

Problem 7.7. Write a program that will show the trajectory of a particle sliding freely on a Hooke plane, but assuming that the plane has a tilt with respect to the vertical, so that one must include the reactive force and its variations with variations in vertical acceleration.

EXERCISES

Exercise 7-1 [lines 30000–30490] *Particle on spheroidal earth*. This program is set up to show the paths of particles on a spheroidal earth. The lefthand sphere is at rest in absolute space, the righthand sphere is shown in a reference frame rotating with angular velocity *W*. Default values of *W*, of *DT*, and of *FACT* (which fixes the scale of the display) and *INC*, the inclination of the spheres in from our point of view are given in line 30090. The latitude *LA(I)* and longitude *LO(I)* have indices 1 to 4 to keep track of which particle trajectory is being computed. *I* = 1 the expert in absolute space, *I* = 2 is the novice in absolute space, *I* = 3 is the expert in the moving reference frame, and *I* = 4 is the novice in the moving reference frame. Particles have different colors, as indicated in the display labels. When a particle gets around on the back of the sphere its color changes. The program asks for you to input the starting latitude by LAT = ?. You enter this latitude in degrees. The program then asks you for an initial starting velocity, with reference to the rotating reference frame: u, v = ?, which means eastward and northward components. Here you enter two numbers, separated by a comma. The rotation rate of the earth is 10, so all the numbers you enter are "normalized to 10." The display then begins.

Line 30160 then converts these two numbers to velocities in the absolute reference frame, in which the calculation is actually performed. The suffix D represents a time derivative. The integration begins in line 30170, using equations in the absolute frame. In line 30220 the relative coordinates are computed. The loop 30230–30340 converts the $LO(I)$ and $LA(I)$ to X, Y, Z coordinates with rotation around the x axis by $SIN(INC)$ and $COS(INC)$ to get the projections correct for the degree of inclination of the display, and then it plots the positions of the four points on the screen [lines 30300–30330] after color choices $C(I)$ are made (lines 30260–30290). Lines 30360–30490 are a subroutine for drawing the coordinates of the two spheres.

Maybe you should start with lat = 45 and u, v = 0, 0. This means that the novice particle is the same as the expert. You will see that he is standing on top of the expert and, because he is plotted after the expert, obscures him. The novice circles the sphere in absolute space at constant latitude, but is stationary in the moving frame. The reason that he can go around the sphere in a latitude circle instead of a great circle is because of the equatorial bulge (you will notice that we keep speaking of a sphere, although it is really a slightly oblate spheroid—too slightly to be visible on the display, and anyway we have by approximation ignored it in the sense described in the text). If you type "s" you can remove the bulge (by setting $W2 = 0$ in line 30180; but this does not reset W, so the moving frame still moves). We use W for Ω in this program. Then you will see that the particles move about in great circles in the absolute reference frame, but in a more elaborate pattern in the moving reference frame. This is of course consonant with our intuition that on a perfect sphere a particle with any initial absolute angular velocity *LOD* would initially slide toward the equator, and that as it did so its angular velocity would become smaller, by conservation of angular momentum.

As we have seen, the Coriolis parameter on a sphere is less strongly a function of latitude near to the poles, so our first experiment could be with lat = 70 and *u*, *v* = 1, 0. This corresponds to a slight push (in relative space) toward the east. Now watch carefully what happens. In the absolute frame the novice outstrips the expert toward the east, but since his *LOD* is too large to be balanced by the bulge (lines 30180) he gets accelerated toward the equator, and begins to mount the equatorial bulge. Now he begins to fall behind the expert in longitude *LO* and then catches up with the expert again. In absolute space he follows an oscillating track, touching the expert somewhat less than two times for each revolution of the expert around the earth. In the moving reference frame the expert is stationary, and the novice moves in an inertial circle. The Coriolis parameter on a rotating sphere is not exactly twice *W*, even at lat = 70, so the circle does not exactly close and its period is not exactly half the rotation period of the earth. When comparing the two reference frames you will notice that the latitude of the novice is always the same in both of them: it is only the longitude in the two systems that is different.

The circle that, in this case, the novice follows appears to be somewhat oblique, but this is because you are looking at the arctic regions rather obliquely. You can get a view more nearly from the pole if you change the inclination *INC* in line 30090. The interpretation of this circle in the rotating frame is that there is a Coriolis force acting to the right of the relative direction of motion, and so the novice goes around in a circle, whose radius depends upon his initial relative velocity. You can try various directions of the initial relative velocities: $u, v = 0, 1; -1, 0; 0, -1$. Each time you will get an approximate circle, executed clockwise, in only a little more than 12 hours. The dynamics of this system is actually computed in the absolute reference frame, but appears to be different in the rotating frame—these are purely kinematic differences due to a simple difference in the time sweep in longitude.

Now let us try a somewhat larger relative velocity, with lat = 70; $u, v = 2, 0$. This gives the novice a larger initial relative velocity to the east. He pulls farther from the expert than before because he is farther from equilibrium of the expert. His inertial circle is twice as large, but has almost the same period (we should expect it to be a little longer period because he goes closer to the equator, where the Coriolis parameter is smaller).

To get a better picture of how the variation of the Coriolis parameter with latitude affects the inertial motions we will now move closer to the equator, for example to 30°N, where the Coriolis parameter has only one half its polar value. Enter lat = 30; $u, v = 0.5, 0$. We see the circle better now because we are looking more directly downward at it. We have chosen only half the starting velocity that we used at 70°N but the circle is visually about the same size. You should think a little about why this is so.

Remaining at 30°N we now try a stronger northward starting velocity on the moving reference sphere: $u, v = 0, 1$. This gives a pretty large circle. In fact it is so large that with constant speed (the speed has to be constant because a normal Coriolis force can do no work on the particle) it is deflected more strongly, and curves more, at high latitude than at low latitude portions of its circuit.

This results in a net westward drift of the particle's orbit. It tends to follow a "flattened corkscrew." Now try initial velocities $u, v = 1, 0; -1, 0; 0, -1$. Does it surprise you that the drift is always to the west?

If you enter even larger relative initial velocities u, v the particle may reach the equator, where the sign of the Coriolis parameter changes sign. Then some pretty trajectories appear. You can find, with a little patience, a value of u that causes the particle to execute a stationary (in the moving frame) figure-eight trajectory, and with somewhat larger values of u you can get snakelike trajectories with the novice moving eastward in relative space.

Here is a special case that sometimes surprises professionals. Try a really large negative v. There now appears to be a Coriolis deflection to the left of the direction of relative motion.

```
30000 'Exer 7-1; particle on spheroid earth **
30010 SCREEN 1: COLOR 0,2: KEY OFF: CLS
30020 'absolute 1 and 2, relative 3 and 4
30021 'expert odd, novice even
30030 PRINT "TWO VIEWS"
30031 PRINT"expert":PRINT"novice"
30040 LOCATE 2,30:PRINT"on back "
30050 PSET(292,12),1
30060 LOCATE 21,6:PRINT"absolute"
30070 LOCATE 21,24:PRINT"relative"
30080 PSET(80,12),2:PSET(80,20),3
30090 PI = 3.14159: W=10: W2 = W^2
30091 DT =.005: FACT = 60:INC=PI/16
30100 GOSUB 30360
30110 LOCATE 23,1 :INPUT "lat";LA
30120 FOR I = 1 TO 4: LA(I)=LA*PI/180
30121 LO(I)=-PI/2    :NEXT
30130 LAD(1)=0: LAD(3)=0
30131 LAD(2)=V:LAD(4)=LAD(2)
30140 LOCATE 23,20:INPUT "u,v";U,V
30150 LAD(1)=0: LAD(3)=0
30151 LAD(2)=V:LAD(4)=LAD(2)
30160 LOD(2)=U/COS(LA(1))+W : LOD(4)=LOD(2)
30170 LODD(2)=2*TAN(LA(2))*LAD(2)*LOD(2)
30178 AAA=SIN(LA(2))*COS(LA(2))
30179 BBB=(W2-(LOD(2))^2)
30180 LADD(2)=AAA*BBB
30190 LOD(2)=LOD(2)+DT*LODD(2)
30191 LAD(2)=LAD(2)+LADD(2)*DT
30200 T=T+DT:A$=INKEY$:IF A$= "s" THEN W2=0
30201 'LOCATE 3,15: PRINT "BULGE GONE":BEEP
30210 LA(2)=LA(2)+DT*LAD(2)
30211 LO(2)=LO(2)+DT*LOD(2):LO(4)=LO(2)-W*T
30220 LA(4)=LA(2): LO(1)=LO(1)+W*DT
30230 FOR I = 1 TO 4: RHO(I)=COS(LA(I))
30240 X(I)=RHO(I)*COS(LO(I))
30241 Y(I)=RHO(I)*SIN(LO(I))*SIN(INC)
```

```
30250 Z(I)=Y(I)+SIN(LA(I))*COS(INC)
30260 IF Y(1)<0 THEN C(1)=2 ELSE C(1)=1
30270 IF Y(2)<0 THEN C(2)=3 ELSE C(2)=1
30280 IF Y(3)<0 THEN C(3)=2 ELSE C(3)=1
30290 IF Y(4)<0 THEN C(4)=3 ELSE C(4)=2
30300 IF I=1 THEN GOTO 30301 ELSE GOTO 30310
30301 PSET(70+FACT*X(1),90-FACT*Z(1)),C(1)
30302 GOTO 30340
30310 IF I=2 THEN GOTO 30311 ELSE GOTO 30320
30311 PSET(70+FACT*X(2),90-FACT*Z(2)),C(2)
30312 GOTO 30340
30320 IF I=3 THEN GOTO 30321 ELSE GOTO 30330
30321 PSET(220+FACT*X(3),90-FACT*Z(3)),C(3)
30322 GOTO 30340
30330 IF I=4 THEN GOTO 30331 ELSE GOTO 30340
30331 PSET(220+FACT*X(4),90-FACT*Z(4)),C(4)
30332 GOTO 30340
30340 NEXT
30350 GOTO 30170
30360 J = INC: ' plot the two spheres "
30370 FOR I = 0 TO 1
30380 CIRCLE (220-150*I,90),60,1,0,2*PI,1
30390 CIRCLE (220-150*I,90),60,1,0,2*PI,SIN(J)
30400 C9=60*SIN(PI/6)*COS(J)
30405 AN=SIN(J)
30410 K9=60*COS(PI/6)
30420 CIRCLE(220-150*I,90-C9),K9,1,0,2*PI,AN
30430 CIRCLE(220-150*I,90+C9),K9,1,0,2*PI,AN
30440 C9=60*SIN(PI/3)*COS(J)
30450 K9=60*COS(PI/3)
30460 CIRCLE(220-150*I,90-C9),K9,1,0,2*PI,AN
30470 CIRCLE(220-150*I,90+C9),K9,1,0,2*PI,AN
30480 NEXT
30490 RETURN
```

Exercise 7-2 [lines 31000−31230] *Equilibrium of a particle at rest on a rotating earth.* This is a display program without operator interaction. It shows a spheroidal earth in absolute space and a particle rotating with it—an expert. The only two forces acting are the gravitational force acting toward the sent of the spheroid, and the reaction of the surface of the spheroid on the particle. The reaction is normal to the spheroid, so it does not line up with the gravity. There is therefore a resultant force (the third line) pointing toward the axis of rotation. It is this unbalanced component of force that accelerates the particle around the earth.

```
31000 'Exercise 7-2; forces on a particle on
31001 'a rotating spheroidal earth
31010 PI=3.14159
31020 KEY OFF: SCREEN 1: COLOR 0,2: CLS
31030 LOCATE 1,17
31031 PRINT "IMBALANCE OF FORCES"
31032 LOCATE 1,37
31033 PRINT" ON A ROTATING SPHEROID"
31040 LOCATE 3,4: PRINT "reactive force"
31050 LOCATE 4,4:PRINT "gravitation"
31060 LOCATE 5,4:PRINT "acceleration"
31070 LINE ( 0 ,20)-( 20,20),3
31080 LINE ( 0 ,28)-( 20,28),2
31090 LINE ( 0 ,36 )-( 20, 36),1
31100 DT =-.21: R0 = .707*175/2
31101 D1=.35:D2=.3:D3=.6
31110 CIRCLE (150, 92),R0 ,1,0,2*PI,.2
31120 X=150+R0*COS(T+PI):Y=.2*R0*SIN(T+PI)+92
31130 T = T+DT
31140 CIRCLE (150,125), 87,1,0,2*PI,.571
31150 CIRCLE(150,125),175/2,1,0,2*PI,.2
31160 CIRCLE(150, 92),R0      ,1,0,2*PI,.2
31170 CIRCLE(150,158),R0      ,1,0,2*PI,.2
31180 LINE(150,78 )-(150,172),1
31185 AAA=92+D3*(Y-92)
31190 LINE(X,Y)-(150+D3*(X-150),AAA),1
31200 LINE(X,Y)-(X+D1*(X-150),Y+D1*(Y-200)),3
31205 AAA=120+D2*(Y-92)
31210 LINE(X,Y)-(150+D2*(X-150),AAA),2
31220 CIRCLE ( X,Y ),2,3: PAINT (X,Y),3
31230 GOTO 31120
```

Exercise 7-3. [lines 32000−32210] *Vertical drop from the Tower of Pisa.* When a particle is dropped from the Tower of Pisa, height $Z = H$ meters, earth's angular frequency *OMEGA*, latitude *TH*, radius of earth *A* with vertical velocity *ZD* (assignable in lines 32020−32030) its velocity is largely vertically downward, and the horizontal components of velocity small. There is, however, a Coriolis force in the eastward momentum equation (6.25) that is proportional to *ZD* and will, with *ZD* < 0 accelerate the particle toward the east. Lines 32040−32080 are simple graphics. The integration of the vertical equation $ZDD = -G$ (32090) and the eastward equation with the approximation that the only Coriolis force that is important is due to *ZD* (32100) is a very simple. Plotting of the position in a vertical/east plane and of several quantities as functions of time, to help in interpretation of the results is done in lines 32120−32150. Intercept 32170 prints out the time and eastward displacement of the particle when it hits the ground ($Z = 0$). Intercept 32180 takes us to subroutine 32200−32210 which reinitializes the computation and shows us the results of another experiment. In the first experiment we dropped the particle from rest at the top of the tower. We got a small eastward displacement due to the earth's rotation. In the second experiment we threw it up fast enough for it to have reached the height of the tower before falling to earth again. Obviously there is a kind of symmetry in the vertical problem. The time elapsed before the particle hits the ground again will be twice that of the first experiment. The particle first goes up and then goes down. Will this reversal of vertical velocity result in a net zero eastward displacement? You really ought to work this out analytically. The large westward displacement in the second experiment comes as something of a surprise.

Try reducing gravity, such as might be proper when considering a neutrally buoyant float falling in the ocean (without drag resistance of course). Try taller towers.

When the towers are built really high—say 6000 kilometers high—then you'd better not use the approximations used above. It will probably be better to think of the whole affair as occurring in absolute space, and the trajectory of the dropped particle as a Keplerian ellipse about the earth's center.

```
32000 'Exercise 7-3 Vertical drop from
32001 'the Tower of Pisa
32010 SCREEN 1: COLOR 0,2: KEY OFF: CLS
32020 G=9.8:H=50:DT=.01:OMEGA=1.4*.0001
32021 A = 6000000!: FAC = 2000!
32030 ZD=0:Z=H:TH=45:PI= 3.14159265#
32031 TH = TH*PI/180
32040 LINE(0,150)-(320,150),1
32050 LINE(100,100-H)-(100,100),1
32060 LINE(0,100)-(300,100),1
32065 LOCATE 1,1:PRINT "side view from south"
32069 LOCATE 13,25
32070 PRINT USING "x(east)*####";FAC
32071 LINE (220,148)-(320,148),2
32072 LINE (220,156)-(320,156),3
32073 LINE (220,164)-(320,164),1
32075 LOCATE 19,25:PRINT"z"
32076 LOCATE 20,25:PRINT"zd"
32077 LOCATE 21,20:PRINT"longitude"
32080 LOCATE 6,13: PRINT "z= 50m"
32090 ZDD= -G : ZD=ZDD*DT+ZD: Z=ZD*DT+Z
32100 PDD =-2*OMEGA*ZD/(A+Z)
32101 PD = PDD*DT+PD: P=PD*DT+P
32110 T = T+DT
32120 PSET( 100+FAC*A*COS(TH)*P,100-Z),3
32130 PSET( 30*T, 150- Z/2),2
32140 PSET( 30*T, 150- ZD   ),3
32150 PSET( 30*T, 150- FAC*P*A*COS(TH)/4),1
32160 LOCATE 1,1: IF FLAG = 1 THEN LOCATE 2,1
32169 AAA=A*COS(TH)*P
32170 IF Z<0 THEN GOTO 32171 ELSE GOTO 32180
32171 PRINT USING "t=##.## sec      ";T
32172 LOCATE 1,17:IF FLAG=1 THEN LOCATE 2,17
32173 PRINT USING "x(east)=##.###m.";AAA
32174 FLAG=1
```

```
32180 IF Z<0 THEN GOSUB 32200
32190 GOTO 32090
32200 ZD=31.26:Z=0:P=0:PD=0:T=0
32210 RETURN
```

SOME FURTHER THOUGHTS ABOUT
THE EXERCISES OF CHAPTER 7

At the end of the exercises of chapter 3 we compared the trajectories that a disturbed (novice) particle describes to that of an expert particle in solid rotation. We found that the novice, if started close to the expert, always remained in its neighborhood. In chapter 4 the novice slowly drifted away from the expert, as a result of the variation of the reaction of the platform on the particle associated with vertical accelerations. When viewed from the rotating axes the paraboloidal dish involves forces other than purely Coriolis ones, and in this respect might be thought of as an imperfect platform.

In the present chapter, the spheroidal platform does indeed imply radial accelerations, and variations in reactive force, but we have replaced it with an approximate sphere, so these radial accelerations of the particle do not appear. Still, when we conduct experiments using exercise 7-1 we discover that novices wander away in earth longitude from the experts. It is interesting to seek some explanation in terms of specially simple cases.

For example, in absolute space the approximate (spherical instead of spheroidal) equation for free motion of a sliding particle in the northward direction is

$$\ddot{\vartheta} + \sin\vartheta\cos\vartheta\,\dot{\phi}^2 = \Omega^2\sin\vartheta\cos\vartheta.$$

The right hand side is the northward component of the reactive force due to the equatorial bulge in absolute space. The expert rides, with $\ddot{\vartheta} = \dot{\vartheta} = 0$ and $\dot{\phi}_e = \Omega$. The halfperiod that the expert takes to go half way around the earth is

$$\tfrac{1}{2}P_e = \pi/\Omega.$$

Suppose in the spirit of the previous discussions we choose as novice a particle starting at rest in absolute space at the latitude of the expert, and of course with no absolute angular momentum at all (Problem 7.2). The appropriate equation of motion is therefore

$$\ddot{\vartheta} = \Omega^2 \sin\vartheta\cos\vartheta = (\Omega^2/2)\sin 2\vartheta.$$

If we write an angle χ that vanishes at the north pole

$$\chi = \pi - 2\vartheta,$$

then this equation is simply

$$\ddot{\chi} = -\Omega^2 \sin\chi.$$

This is our familiar friend the equation for a spherical pendulum

$$\ddot{\chi} = -\frac{g}{e}\sin\chi,$$

which can be shown to have the period

$$\frac{1}{2}P = \frac{\pi}{\Omega}\left(1 + \frac{1}{4}\sin^2\frac{\chi_0}{2} + \cdots\right),$$

where χ_0 is the amplitude of the swing. Evidently, the novice will not get across the pole quickly enough to meet the expert at the other side of the earth, and so it will slowly drift behind it in longitude. In the system moving with the earth, the novice does loops like inertial circles, but they are not closed, and slowly drift westward. You can compute a clear demonstration using exercise 7-1 if you change, in line 30120, $LO(I)$ to $LO(I) = -PI^*(.55)$.

The reason for doing this is to rotate the display a little in longitude so that you can observe the novice's path across the pole from a side-angle, and thus see both ends of its path. Then one enters Lat = 30; u, v = −8.660, 0. (Remember that the u, v are velocities relative to the rotating earth, and that the equatorial velocity was scaled to 10 arbitrary units.) This means that the novice has no absolute velocity at its starting point. You can see it slide toward the pole, cross it, and reach the other extreme of its swing across the pole too late to meet the expert who started at the same place. The apparent westward drift is seen in the righthand panel.

There are many other types of motion that could be discussed as special cases. One of the simplest is to consider, in the moving reference frame, the motion of a novice close to the equator, and whose initial relative velocity and position coincide with an expert who appears to be stationary in the earth's frame. The relative equations are

$$\ddot{\vartheta} + \frac{\mu^2}{a^4\cos^4\vartheta}\sin\vartheta\cos\vartheta = \Omega^2\sin\vartheta\cos\vartheta.$$

For small latitudes this reduces to a simple harmonic equation for oscillation across the equator. It is a simple problem to find the analytical expression for the wavelength of the path described by a novice particle started off eastward close to the equator. It is also interesting to enquire what happens to a particle that started westward. You can try these two cases with exercise 7.1 by setting Lat = 3 and u, v = 1, 0 or −1, 0.

CHAPTER VIII

Forced motion

8.1 REAL FORCES RELATIVE TO THE ROTATING SYSTEM

Now we want to discuss two types of forced motion. In the first type the force will be specified in terms of the coordinates of the rotating reference frame. Here we might think of a spring which ties our novice particle to a point fixed in the rotating coordinate system. Or we might have in mind a local hillock on the topography of the rotating frame, capable of producing a horizontal component of reaction to force the novice downhill. A geographically fixed atmospheric low pressure region (in the absence of fluid drag) could tend to draw a novice toward its center.

The second type of force is one which comes from an interaction of other particles that are also free to move on the platform but whose coordinates are given in the moving frame. This is much more like a thin layer of gas or fluid covering the surface of the platform. If two novices get close to each other they may attract or repel each other along the lines joining them with a force proportional to some power of their distances of mutual separation. This type of force is a primitive representation of the forces between particles of air or water in the atmosphere and ocean. But we do not want to intrude into the large subject of hydrodynamics because we risk losing our focus on the Coriolis force. We therefore will eschew a digression into the physics of continua. Our aim is simply to discover some of the phenomena that can occur when real forces in the rotating reference frame coexist with the virtual Coriolis force.

8.2 BALANCES AMONG TERMS

The introduction of an additional real force in the rotating frame means that now the equations of motion in the rotating frame have three terms: (a) acceleration in relative coordinates, (b) Coriolis forces, and (c) real forces. Up until now we have considered only balances between (a) and (b). The motions that we observed with this balance of (a) and (b), some complicated indeed, are all called—in oceanographic and meteorological parlance—*inertial* motions. Motions resulting from a balance between forces (b) and (c) can be called *geostrophic* motions. With the balance between (a) and (c) we are dealing with the phenomena of the everyday workplace: the earth's rotation is not important, and so the phenomenon is effectively occurring in absolute space.

8.3 RESPONSE OF A PARTICLE TO A FORCE OF THE FIRST TYPE

Imagine that a particle is at rest in the rotating frame, and that a force is applied to the particle. We pose the problem of discovering how the particle will respond. Will the work done on the particle go mostly into energy of inertial motions, or into energy of geostrophic ones, or into energy of the everyday workplace?

The simplest type of force in the rotating Hooke plane is a constant force, say F_y = constant that is turned on suddenly at time $t = 0$ and remains constant (with reference to the rotating plane) thereafter. The reason that we have to make that parenthetical qualification is that our force has its direction fixed in the rotating plane, like the trade wind-stresses on the rotating earth, for example. Of course this "constant force" must actually be rotating in absolute space, so here some of the advantages of thinking in terms of rotating reference systems already begin to emerge. If we are formulating a problem where our positions, boundaries, force fields, etc. are most naturally and easily referred to the rotating frame, and the frame itself is in a rotating dynamical equilibrium, then it is easier to set up the dynamics relative to the rotating frame, at the small cost of introducing Coriolis forces. Proceeding with our example, the dynamical equations for the simple forced system we envisage is

$$\ddot{x}' = 2\Omega\dot{y}'$$
$$\ddot{y}' = -2\Omega\dot{x}' + F_{y'} \qquad (8.1)$$

where the terms on the left are the x, y accelerations relative to the moving reference frame, the first terms on the right are the Coriolis forces (you will notice they vanish if the particle is at relative rest) and the impulsively turned on force $F_{y'}$ is also there, fixed with respect to the rotating platform. The equations are simple, but remember that in back of them there is the implicit existence of a dynamical equilibrium of solid rotation with the Hooke springs that involve real axially directed forces on all particles (including those that make up our imaginary reference frame and guarantee its solid rotation). Note also the disappearance of the other virtual forces—except the Coriolis forces—from the right hand side.

Assuming that $x' = y' = \dot{x}' = \dot{y}' = 0$ at $t = 0$ the solution is a combination of a steady $\dot{x}' = F_{y'}/2\Omega$ and a free circular inertial motion. We would like to call the steady part a geostrophic motion (in geophysical literature this term is often reserved for flows whose Coriolis force balances a horizontal pressure gradient, but here we do not have fluids, just particles). The solution of equation (8.1) is

$$x' = \frac{F_{y'}}{4\Omega^2}(2\Omega t - \sin(2\Omega t)),$$

and

$$y' = \frac{F_{y'}}{4\Omega^2}(1 - \cos(2\Omega t)).$$

The average kinetic energy (in the rotating frame) of the geostrophic motion of the particle is

$$\frac{1}{2\pi}\int_0^{2\pi} \frac{F_{y'}}{2\Omega} d(2\Omega t),$$

and of the inertial motion is

$$\frac{1}{2\pi}\int_0^{2\pi} \frac{F_{y'}}{2\Omega}(\sin^2(2\Omega t) + \cos^2(2\Omega t))\,d(2\Omega t),$$

so there is equipartition of energy between the two types of motion. If the force $F_{y'}$ is turned on slowly instead of abruptly, less energy will go into the inertial type of motion. It is easy to see that if half of the force is turned on at $t = 0$ and the second half turned on at the time corresponding to one half the inertial period, the net contribution to the inertial oscillation will be entirely canceled out, and only steady "geostrophic" drift of the particle to the right of the applied force will remain. The force balance in the rotating reference frame will then be extremely simple. The applied force $F_{y'}$ is to the north, and the Coriolis force that balances it is to the south. The Coriolis force is toward the south because the particle is moving steadily to the east [for $\Omega > 0$]. As we have seen this is a nice balance, it takes a delicate control at the beginning of setting up the force to avoid oscillations about the drift.

If the force field **F** varies with time periodically the partition of energy of the forced motion will favor geostrophy when the period of the forcing is longer than that of the Coriolis period. This is generally true in the large, slowly moving ocean currents and atmospheric winds.

Exercise 8-3 includes an example of two mutually attracting particles. Exercise 8-4 enables you to explore the case of a small spring acting upon a particle in a rotating reference frame. Depending upon the strength of the spring the balance of terms in the equations in the rotating reference frame will be different, and we can get some insight on why for certain set-ups an experiment will be almost unaffected (in the rotating frame) by the earth's rotation, whereas in others the particle will move with the full balance being between the Coriolis force and spring force on the right hand side. The latter type of motion might be called geostrophic. Of course we are not dealing explicitly with a fluid or gas layer on the earth, so we don't have physical concepts like pressure and mass continuity as in hydrodynamical language. So it is possible that some readers may take exception to this use of the word geostrophic in connection with the dynamics of a particle.

Geostrophic motions are of special interest to meteorologists and oceanographers. At first they seem rather mysterious. However, when we think of the slow motions of the atmosphere and ocean that make up winds and ocean currents we realize that they are very small local perturbations to the uniform solid rotation of the earth. Because of the equatorial bulge, the forces with which they are associated are much smaller than the poleward pressure gradients. The bulges in the ocean surface that provide the horizontal forces that balance these slight perturbations in the otherwise solid rotation of the ocean amount to about 2 meters water pressure across the largest of ocean currents; and those in the atmosphere are only a tenth of that. The picture to keep in mind is that the earth and air and sea are all circling the axis of the earth at very nearly the same angular velocity, that there are forces deriving from gravitational attraction and mutual reactions of all their constituent parts that are out of balance to maintain that rotational equilibrium, and that the "out of balance" that derives from the slight differential rotation of the winds and currents are small perturbations of, but part of, the overall equilibrium.

EXERCISES

Exercise 8-1 [lines 35000–35140] *Sudden force in the relative frame.* This program models the effect of a sudden force fixed in amplitude and direction relative to the moving reference frame on a particle initially at rest in that frame. The problem is done with the simple Hooke platform of chapter 3, and the coordinates are the relative ones. The program uses X and Y, but we really mean primed quantities, MX, MY, which, for convenience's sake, we leave out. There is nothing to manipulate in this program. For a short time after it starts we see a resting particle. Then a force *FO* turns on in the integration loop in line 35080. The direction of the force is indicated by the arrow that turns on when the force does. The particle initially begins to move in the direction of the force, but the Coriolis force deflects it to the right of its motion and it turns toward the *x* direction, and then in the negative *y* direction in an inertial oscillation. There is a constant drift of the inertial circle toward the *x* direction and this is what balances the force *FO*. We can speak of the motion then having two parts, an inertial one where accelerations *XDD, YDD* are balanced by part of the Coriolis force, and another part of the motion in which the steady force *FO* is balanced by the remainder of the Coriolis force. This has an important example in the ocean where (theoretically) the mean drift of a surface layer in the ocean is to the right (to the left in the southern hemisphere where the sign of the Coriolis parameter is opposite) of the applied wind-stress.

```
35000 'Exercise 8-1; fixed force on ********
35001 'relative frame
35010 KEY OFF: SCREEN 1: COLOR 0,2: CLS
35020 X = 1 :Y =.1:LOCATE 2,2: PRINT "my"
35030 F = 1 : DT = .1: FACT = 5
35040 LINE(0,0)-(0,199),1
35041 LOCATE 23,35: PRINT "mx"
35050 LINE(0,199)-(300,199),1
35060 W = 10 ' Coriolis parameter 2*omega
35070 XDD=YD: XD=XDD*DT+XD:X=XD*DT+X
35080 YDD=-XD+FO: YD=YDD*DT+YD:Y=YD*DT+Y
35090 PSET(FACT*X,100-FACT*Y),3
35100 COUNT = COUNT +1
35110 IF COUNT > 40 THEN FO=F
35120 AX= 200: AY = 100:LX=0: LY = 50  *F
35121 AL=.4:LH=4:CA=2
35130 IF COUNT > 40 THEN GOSUB 60000
35140 GOTO 35070
```

Exercise 8-2 [lines 36000−36430] *Particle subject to a force directed toward a point fixed on a moving frame.* There are two options in this program. Option 1 turns the force on fully at the initial moment $t = 0$. Option 2 turns the force on gradually as a linear function of time until it reaches its full strength at the time when the platform has made one revolution, and thereafter holds the force constant. The location of the fixed point could be an expert at rest in the moving frame, but no back reaction of the moving particle upon the expert is allowed [lines 36210−36230]. The display shows both frames of reference, absolute and relative. The circles are constant radii. The expert (force center) is index $I = 0$, the novice is index $I = 1$. They both start in equilibrium [lines 36070−36080] and would both be expert were it not for the force of $I = 0$ on $I = 1$. The calculation is done in rectilinear coordinates in absolute space. The main integrations are lines 36240−36290 inside a loop that does them both. The quantity $D(I)$ is the distance between the two particles [36160]. *EP* [36050] in the constant amplitude of the force, converted to instanteous values by lines 36210−36230. The law of the force is $EP*D(I)\hat{\ }M$, the M power has been set for default at $M = -2$ (gravitational law) in line 36060. You can experiment with other power laws. The angle AL is the angle between the x axis and the line joining the moving particle to the center of attraction (expert).

Plotting is done in the loop 36370−36420: first the absolute frame [36380] and then after rotation of coordinates to remove the rotation of the frame [36390−36400] the relative plot is made [36410].

What we see in this program is how the novice responds to the suddenly or gradually applied force. With the sudden force there is a strong inertial oscillation in addition to a circular geostrophic drift around the force center. With the gradually applied force the inertial oscillation is largely eliminated. The time to execute an inertial oscillation is always half the rotation period of the frame, but the time to go around the center of attraction geostrophically depends upon the amplitude of the force at the mean radius of oscillation. Since this radius also depends upon the sign of the force the geostrophic period can also be different for orbits commenced with all other variables the same except the sign of the central force. In a very crude way we are modeling the dynamics of a particle subject to forces analogous to those that a radially directed horizontal pressure gradient, or the slope of a hill, might produce on a particle in the system. Perhaps we can think of atmospheric highs and lows, depending upon whether the force is directed outward from or inward toward the center. Remember, the force center is supposed to be fixed to the rotating reference frame, as would be the case of a stationary atmospheric pressure feature, or certainly a little hill.

```
36000 'Exercise 8-2: Particle subject to a***
36001 'force exerted by an expert
36005 SCREEN 1: COLOR 0,2: KEY OFF: CLS
36010 LOCATE 4,1
36011 PRINT "Opt 0: full force on at t=0"
36020 LOCATE 5,1
36021 PRINT "Opt 1: force with time-ramp "
36030 LOCATE 7,4: INPUT "Opt=";OPT
36040 CLS
36050 NUM = 1: DT=.03: C = 1
36051 EP=.1 : PI = 3.14159365#
36060 N = 1: M=-2
36070 X(0)=1: Y(0)=0 : X(1)=.5: Y(1)=0
36080 YD(0)= 1: XD(0)=0: XD(1)=0: YD(1)=.5
36090  CIRCLE ( 75 ,100),50,1
36100  CIRCLE (225 ,100),50,1
36110  PSET   ( 75 ,100),  1
36120  PSET   (225 ,100),  1
36130 LOCATE 23,5
36131 PRINT " absolute        relative"
36140 FOR I = 0 TO NUM
36150 R(I)=SQR(X(I)^2+Y(I)^2)
36160 D(I)=SQR((X(0)-X(I))^2+(Y(0)-Y(I))^2)
36170 NEXT
36180 LOCATE 2,12
36181 PRINT USING "revs = ###.###";T/(2*PI)
36190 T = T+DT
36200 FOR I = 0 TO NUM
36210 IF T < 2*PI GOTO 36211 ELSE GOTO 36212
36211 EPO=EP*T/(2*PI):GOTO 36220
36212 EPO=EP:GOTO 36220
36220 IF OPT = 0 THEN EPO = EP
36230 IF I = 0 THEN E = 0 ELSE E = EPO
36240 AAA=E*D(1)^M*COS(AL(I))
36241 XDD(I)=-C*R(I)^N*COS(TH(I))+AAA
```

```
36250 XD(I)=XDD(I)*DT+XD(I)
36260 X(I)=X(I)+XD(I)*DT
36270 AAA=E*D(1)^M*SIN(AL(I))
36271 YDD(I)=-C*R(I)^N*SIN(TH(I))+AAA
36280 YD(I)=YDD(I)*DT+YD(I)
36290 Y(I)=Y(I)+YD(I)*DT
36300 NEXT
36310 FOR I = 0 TO NUM
36320 TH(I)=ATN( Y(I)/X(I))
36321 IF X(I)<0 THEN TH(I)=PI+TH(I)
36330 NEXT
36340 FOR I = 1 TO NUM
36350 AL(I)=ATN((Y(0)-Y(I))/(X(0)-X(I)))
36351 IF (X(0)-X(I))<0 THEN AL(I)=AL(I)+PI
36360 NEXT
36370 FOR I = 0 TO NUM
36380 PSET( 75  +50*X(I), 100-50*Y(I)), I+2
36390 XP(I)=X(I)*COS(TH(0))+Y(I)*SIN(TH(0))
36400 YP(I)=-X(I)*SIN(TH(0))+Y(I)*COS(TH(0))
36410 PSET( 225 +50*XP(I), 100-50*YP(I)), I+2
36420  NEXT
36430 GOTO 36140
```

Exercise 8-3 [lines 37000−37280] *Geostrophic adjustment for two interacting particles on plane.* This program has two particles a distance *R* apart that have an attraction or repulsion between them that is a function of the distance −*K*R^N*. You can set *K* and *N* in line 37040. The force turns on suddenly. The program is entirely written in the rotating reference frame (but you will notice that for simplicity the prefix *M* does not appear in the program). You are reminded that the stars move with respect to you by the rather time-consuming star display [subroutine 61000 + ···] you can remove line 37210 to get rid of the star. The integration loop starting at line 37080 is the simple set of terms discussed in the equations (8.1) of the text.

```
37000 'Exercise 8-3;Two interacting particles
37001 ' entirely written on the rotating
37002 ' reference frame
37020 SCREEN 1: COLOR 0,2: KEY OFF: CLS
37030 LOCATE 1,1
37031 PRINT "in rotating reference frame"
37040 DT = .05 :C0 = 1: BETA =0
37041 K=-100:N=1: FRICT=0
37046 LINE (0,100)-(300,100),1
37047 LINE (150,0)-(150,200),1
37050 X(1)=20:X(2)=-20:Y(1)=0:Y(2)=0
37080 FOR I = 1 TO 2  : FOR J = 1 TO 2
37090 C =C0+BETA*Y(I)
37100 IF I = J THEN GOTO 37160
37110 R = SQR((X(I)-X(J))^2+(Y(I)-Y(J))^2)
37120 AAA=(R^(1+N))-FRICT*XD(I)
37121 XDD(I)=C*YD(I)+K*(X(I)-X(J))/AAA
37130 BBB=(R^(1+N))-FRICT*YD(I)
37131 YDD(I)=-C*XD(I)+K*(Y(I)-Y(J))/BBB
37140 XD(I)=XD(I)+DT*XDD(I)
37141 YD(I)=YD(I)+YDD(I)*DT
37150 X(I)=DT*XD(I)+X(I):Y(I)=Y(I)+DT*YD(I)
37160 NEXT:NEXT
37180 T = T+DT
37200 XS=150+92*COS(-T/2):YS=100-92*SIN(-T/2)
37210 GOSUB 61000
37220 XLS=XS:YLS=YS
37230 FOR I = 1 TO 2
37240 PSET (150+XL(I),100-YL(I)),1
37250 PSET (150+X(I),100-Y(I)),2
37260 XL(I)=X(I):YL(I)=Y(I)
37270 NEXT
37280 GOTO 37080
```

Exercise 8-4 [lines 38000–38150] *A particle on the moving frame attached to a point fixed on the frame by an additional Hooke spring.* This little program enables you to make some experiments easily in changing the spring constant K of the little extra spring that acts on the particle in the rotating frame. The Coriolis parameter is set as $C = 1$, and you may think of the value of K (choose between .1 and 10, a range that lets you have a weak or strong spring) as being normalized to the rotational frequency of the frame, $C/2$. When the spring is strong, the particle oscillates quickly across the point of attraction like an ordinary pendulum in a nonrotating reference frame, but it does feel the Coriolis force to some extent and the vertical plane of its oscillation rotates in the apparent direction of the stars. If you want to choose $K > 10$ it is a good idea to diminish the integration step DT in line 38020, to avoid rapidly exploding integration errors. Then of course you will be in the ordinary work-a-day physical world: you will hardly notice the effect of the Coriolis force (a Foucault pendulum works very much like this). The balance in the dynamical equations on the rotating reference frame is between the accelerations on the left-hand side and the spring force on the right-hand side, the Coriolis force being quite small. On the other hand, if you set $K = 1$ you will have a more equal distribution of all three terms acting, and there will be a star-shaped trajectory that is a combination of inertial and geostrophic motions. Finally, if you choose K to be small the natural frequency of the little oscillator will be much less than that of the Coriolis frequency C, and the balance will be mostly between the Coriolis force and the spring force, the left-hand side accelerations being very small. The motion is then a slightly serrated circle—whose sense of rotation depends upon the sign of K—nearly pure geostrophic balance. Inertial circles always (in the northern hemisphere) appear to go around clockwise; geostrophic circles can go in either direction depending upon whether a central force is attractive or repulsive. In this particular exercise it is attractive, so the geostrophic circle is counterclockwise.

```
38000 ' Exercise 8-4; coriolis and spring***
38001 ' on particle
38020 SCREEN 1: KEY OFF : COLOR 0,2: CLS
38030 LOCATE 20,3: INPUT " k = ? (.2 to 10)";K
38040 DT = .02: C =1 : X = 1: Y=1 : K2=K^2
38050 PSET(150,100),2
38060 T = T+DT
38070 XDD =   C*YD-K2*(X-X0)
38080 XD= XDD*DT+XD  : X = XD*DT+X
38090 YDD = -C*XD-K2*(Y-Y0)
38100 YD= YDD*DT+YD:  Y = Y+YD*DT
38110 PSET(150+50*X,100-50*Y),3
38120 XS=150+80*COS(T/2):YS=100-80*SIN(-T/2)
38130 GOSUB 61000
38140 XLS=XS:YLS=YS
38150 GOTO 38060
```

CHAPTER IX

Refining the earth's platform

9.1 DEFICIENCIES OF THE HUYGENS SPHEROID

In chapter 7 we constructed a spheroidal rotating platform with a central gravitational attraction that could support a population of sliding particles in solid rotation about the axis of rotation—all going around with the same period. The equations of motion for a particle sliding on this platform relative to the sedentary cloud (relative to a frame rotating with this unique period) were then written down in spherical coordinates, with the constant radius a replacing the real radius r, a conventional approximation often adopted for problems of motion on rotating spheres. The difference between r and a was small compared to a itself.

There is a difficulty with this platform that at first seems alarming. From equation (7.4) we found that the excess of equatorial radius over the polar radius is $\frac{1}{2}\Omega^2 a^2/g$ and that this amounts to about 11 kilometers. Various geodetical determinations of this excess put it at about 21 kilometers. This is a marked discrepancy. It arises mostly from a deficiency in the way we calculated the direction of gravitational attraction when constructing the platform. If the earth is a spheroid instead of a sphere (which we assumed implicitly) then the direction of gravitation is not strictly toward the geometrical center. In fact, all who have gone through the derivation of the proof—by three dimensional integration of all the inverse square attractions of the elements inside it—that the net attraction of all the particles in a sphere of uniform density upon a particle exterior to it is equivalent to that obtained by gathering all the mass at the center of the sphere, will regard the result as something of a miracle. They will also recognize that it cannot be a general feature for bodies of less special shape. But of course, if you have to know the shape of the platform in order to compute it there is obviously more involved in the physics than our simple construction of the Huygens platform in chapter 7. Determination of the shape of the rotating earth and of the form of the gravitational force field are not separate problems: they are coupled. Let us therefore reexamine what we did in chapter 7 in a slightly different way.

9.2 COMBINED CENTRIFUGAL AND GRAVITATIONAL POTENTIALS

Suppose that in the absolute x, y, z space we have a single particle of unit mass circling the z axis with angular frequency Ω. No matter where the particle is, it circles the axis at the same frequency. It is therefore always subject to a centrifugal force with components $\Omega^2 x$, $\Omega^2 y$, 0. At the origin we place a center of gravitational attraction, strong enough so that at radial distance a the gravitational attraction on the particle is g. Therefore the gravitational attraction on the particle at other radii is ga^2/r^2 and directed toward the origin. Of course if that strong attractor at the origin has a finite spherical size, we have to be careful to stay outside of it, otherwise the gravitational attraction on our particle will be different. We introduce a here only to define a magnitude for g.

Each of these two forces, the centrifugal and the gravitational, can be expressed as the gradient of a potential (a function of position only). Writing Φ_c for the potential of the centrifugal force, and Φ_g for the potential of the gravitational attraction these are

$$\Phi_c = \frac{1}{2}\Omega^2(x^2 + y^2), \qquad \Phi_g = \frac{ga^2}{r}. \tag{9.1}$$

To find the magnitude and direction of both forces combined we need only to perform the gradient operation on Φ the sum of the two potentials.

Figure 9.1 shows contours of Φ in the x, z plane. The contribution of the centrifugal force is stronger relative to gravitation in the figure than on the earth in order to make the ellipse-like form of the contours around the center more visible to the eye. Although the figure shows only the x-z plane, we must think of these contours as defining surfaces of revolution around the z axis. The net force on the particle is normal to these contours of equal

Figure 9.1 [*opposite page*]. This figure shows the curves in the x, z plane that are intersections of the surfaces of revolution (about the z axis) that represent equipotentials of the combined forces of centrifugal force from the axis due to rotation and center-directed gravitational force. The arrows indicate the direction of the force in several parts of the field. The contours become so crowded near the origin that we have left a blank circle there. The ratio of amplitude of centrifugal force to gravitational force is larger in this figure than in the real rotating earth system, so that the ellipsoidal form of the equipotentials at small distance from the origin will be clearly visible.

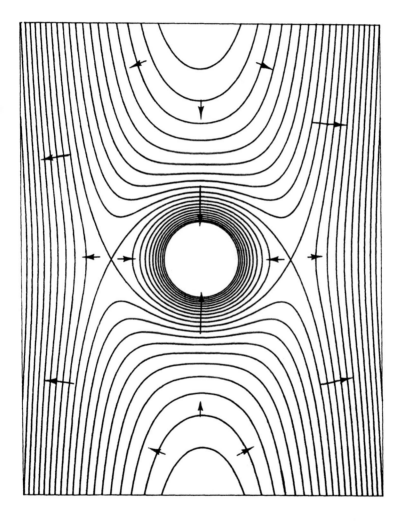

potential (equipotential surfaces), and proportional to the closeness of their spacing, as indicated by a few force arrows on the figure. Exercise 9-1 shows each of the potential fields separately and then shows the sum.

9.3 THE CONCEPT OF A PLATFORM AS AN EQUIPOTENTIAL SURFACE

It is important in looking at these fields to realize that our particle cannot actually continue to circle the axis with frequency Ω in the presence of this force field (except in the one circle where the force vanishes—it passes through those two saddle points on the x axis where centrifugal force exactly balances gravitational attraction—and is the home of geostationary satellites). For the particle to circle the axis elsewhere the force associated with the combined potential Φ must be balanced by something else.

If we could build a massless solid slippery surface to coincide with any one of the equipotential surfaces, then we would have a platform to provide a normal reactive force on the particles and permit them to circle the axis at the angular frequency Ω. For example, up near the top of figure 9.1, where the gravitational attraction is nearly in the $-z$ direction, there are some nearly paraboloidal surfaces reminiscent of our paraboloid of chapter 4. Further down, there are some regions of nearly flat surface, where axial components of the gravitation imitate the Hooke springs of chapter 3. Then there are some strange surfaces that look like the bottom of a champagne bottle, and finally, near the center of attraction at the origin, there are some oblate spheroidal surfaces such as contrived in chapter 7. We chose one of these surfaces tangent to the spherical earth at the pole, at radius a, where

Φ (pole) $= ga^2/a$. Then we went along this equipotential surface to its intersection with the equatorial plane $z = 0$, where the radius was no longer at $r = a$ but determined from $\Phi(\text{pole}) = \Omega^2 r^2/2 + ga^2/r$. This radius

$$r \cong a\left(1 + \frac{1}{2}\Omega^2 a/g\right)$$

was the equatorial bulge of the approximate Huygens platform described in chapter 7. But it lies above the sphere itself, so that platform has to be envisaged as a massless platform raised above the earth 11 kilometers at the equator.

So, why don't we think of the earth's surface as coinciding with the equipotential surface? To accomplish this rigorously the gravitational part of the combined potential must be modified to correspond to that appropriate for the attractive force of a spheroid, differing from that of a sphere, and with an attractive force not directly pointing toward the origin (center of mass of the attracting body). In the sketch that follows we make use of the theory given by the eighteenth-century mathematician Maclaurin.

9.4 MACLAURIN'S ELLIPSOID

Consider an oblate ellipsoid of uniform density bounded by the surface

$$\frac{x^2 + y^2}{a^2} + \frac{z^2}{c^2} = 1. \tag{9.2}$$

We define the polar radius as c, and equatorial radius as a. The eccentricity $e^2 = 1 - c^2/a^2$ so $c^2 = a^2(1 - e^2)$. For this surface to be an equilibrium surface we need to have the combined effect of the gravitational attraction and the centrifugal acceleration produce a resultant force (on a particle on the surface) that is normal to the surface. This is just like the problem illustrated in Figure 7.1, except that now the direction of gravitational attraction is *not* assumed to be toward the center.

To begin, we need to find the gravitational attraction for the ellipsoid. Instead of deriving this potential, we will state the answer and demonstrate its correctness. If the potential is Φ_g the force it produces is $F = \nabla\Phi_g$; that is, the components of the force are

$$F_x = \frac{\partial \Phi_g}{\partial x}, \qquad F_y = \frac{\partial \Phi_g}{\partial y}, \qquad F_z = \frac{\partial \Phi_g}{\partial z}.$$

The requirement that Φ_g be the gravitational potential for a given mass distribution of density ρ (in our case a constant inside the ellipsoid and zero outside it) is

$$\nabla^2 \Phi_g \equiv \left(\frac{\partial^2}{\partial x^2} + \frac{\partial^2}{\partial y^2} + \frac{\partial^2}{\partial z^2} \right) \Phi_g = -4\pi G \rho \qquad (9.3)$$

where G is the gravitational constant. The potential for the oblate ellipsoid is (O. D. Kellogg, 1953. *Foundations of Potential Theory*, New York, Dover Publ.)

$$\Phi_g(x, y, z) = \pi a^2 c G \rho \int_\lambda^\infty \left[1 - \frac{x^2 + y^2}{a^2 + s} - \frac{z^2}{c^2 + s} \right]$$
$$\times \frac{ds}{(a^2 + s)\sqrt{c^2 + s}} \qquad (9.4)$$

where λ is the larger root of

$$\frac{x^2 + y^2}{a^2 + \lambda} + \frac{z^2}{c^2 + \lambda} = 1$$

if we are outside the ellipsoid, and $\lambda = 0$ if we are inside it. This expression involves three integrals:

$$I_0 = \int_\lambda^\infty \frac{ds}{(a^2 + s)\sqrt{c^2 + s}},$$

$$I_1 = \int_\lambda^\infty \frac{ds}{(a^2 + s^2)^2 \sqrt{c^2 + s}},$$

and

$$I_2 = \int_\lambda^\infty \frac{ds}{(a^2 + s)(c^2 + s)\sqrt{c^2 + s}}.$$

So, we can write equation (9.4) as

$$\Phi_g = \pi a^2 c G\rho (I_0 - (x^2 + y^2)I_1 - z^2 I_2). \qquad (9.4')$$

Begin with I_2, and noting that

$$\frac{1}{(a^2 + s)(c^2 + s)} = \frac{1}{(a^2 - c^2)}\left[\frac{1}{c^2 + s} - \frac{1}{a^2 + s}\right],$$

so that

$$I_2 = \frac{1}{(a^2 - c^2)} \int_\lambda^\infty \left(\frac{1}{c^2 + s} - \frac{1}{a^2 + s}\right) \frac{ds}{\sqrt{c^2 + s}}$$

$$= \frac{1}{(a^2 - c^2)}\left(\frac{2}{\sqrt{c^2 + \lambda}} - I_0\right). \qquad (9.5)$$

For I_1 an integration by parts gives

$$I_1 = -\frac{1}{(a^2+s)} \frac{1}{\sqrt{c^2+s}} \Big|_\lambda^\infty - \frac{1}{2} \int_\lambda^\infty \frac{1}{(a^2+s)(c^2+s)} \frac{ds}{\sqrt{c^2+s}}.$$

But the last integral is I_2 again, so

$$I_1 = \frac{1}{(a^2+\lambda)} \frac{1}{\sqrt{c^2+\lambda}} - \frac{I_2}{2}, \qquad (9.6)$$

or

$$I_1 = \frac{1}{(a^2+\lambda)\sqrt{c^2+\lambda}}$$
$$- \frac{1}{(a^2-c^2)\sqrt{c^2+\lambda}} + \frac{I_0}{2(a^2-c^2)} \qquad (9.7)$$

Finally, to evaluate I_0 write $r^2 = c^2 + s$, $r_0^2 = c^2 + \lambda$, $ds = 2r\,dr$ and

$$I_0 = \int_\lambda^\infty \frac{ds}{(a^2+s)\sqrt{c^2+s}} = \int_{r_0}^\infty \frac{2\,dr}{a^2-c^2+r^2}$$

$$= \frac{2}{\sqrt{a^2-c^2}} \left[\frac{\pi}{2} - \tan^{-1} \sqrt{\frac{c^2+\lambda}{a^2-c^2}} \right]$$

$$= \frac{2}{\sqrt{a^2-c^2}} \sin^{-1} \left(\sqrt{\frac{a^2-c^2}{a^2+\lambda}} \right) \qquad (9.8)$$

In particular, if $\lambda = 0$ (inside the ellipsoid), $I_0 = 2\sin^{-1}e/ae$.

Before we prove that $\Phi_g(x, y, z)$ is the desired potential for the oblate ellipsoid, let us do a simple case to see the structure of the integral in equation (9.4), and in particular distinguish between the inside and outside of the ellipsoid. For simplicity consider a sphere. Then the formula is

$$\Phi(x, y, z) = \int_\lambda^\infty \left(1 - \frac{x^2 + y^2 + z^2}{a^2 + s}\right) \frac{ds}{(a^2 + s)^{3/2}}$$

with $\lambda = 0$ if we are inside the sphere, and

$$\lambda = x^2 + y^2 + z^2 - a^2$$

if we are outside it. Now

$$\int_\lambda^\infty \frac{ds}{(a^2 + s)^{3/2}} = \frac{2}{(a^2 + \lambda)^{1/2}},$$

and

$$\int_\lambda^\infty \frac{ds}{(a^2 + s)^{5/2}} = \frac{2}{3} \frac{1}{(a^2 + \lambda)^{3/2}},$$

so we have

$$\Phi = \frac{2\pi a^3 G\rho}{(a^2 + \lambda)^{1/2}} \left[1 - \frac{x^2 + y^2 + z^2}{3(a^2 + \lambda)}\right].$$

If we are inside the sphere, $\lambda = 0$ so

$$\Phi = 2\pi a^2 G\rho \left[1 - \frac{x^2 + y^2 + z^2}{3a^2}\right].$$

If we are outside the sphere, $\lambda = x^2 + y^2 + z^2 - a^2$ so

$$\Phi = \frac{4}{3}\pi a^3 G\rho \frac{1}{\sqrt{x^2 + y^2 + z^2}}.$$

Since the mass of the sphere is $M = 4\pi a^3 \rho/3$, outside the sphere we have $\Phi = MG/R$ exactly as expected. At the surface $R = a$ both values agree. Furthermore it is easy to check that outside the sphere $\nabla^2\Phi = 0$ and inside it $\nabla^2\Phi = -4\pi\rho$. The only feature of the integral that differs between the inside and outside, and therefore accounts for the discontinuous change in $\nabla^2\Phi$ is that the lower limit $\lambda = 0$ if we are inside, but depends upon x, y and z if we are outside.

Returning to the ellipsoid, on the inside

$$\Phi_g(x,y,z) = \pi a^2 c G\rho \int_0^\infty \left[1 - \frac{x^2+y^2}{a^2+s} - \frac{z^2}{c^2+s} \right]$$
$$\times \frac{ds}{(a^2+s)\sqrt{c^2+s}}.$$

Performing the indicated differentiation with respect to x, y, and z gives

$$\nabla^2\Phi_g = \pi a^2 c G\rho \int_0^\infty \left[-\frac{4}{a^2+s} - \frac{2}{c^2+s} \right] \frac{ds}{(a^2+s)\sqrt{c^2+s}}$$
$$= -2\pi a^2 c G\rho[2I_1 + I_2]. \tag{9.9}$$

Now from equation (9.6) we see that when $\lambda = 0$,

$$2I_1 + I_2 = 2/a^2c,$$

so indeed $\nabla^2\Phi = -4\pi G\rho$ as long as x, y, and z are inside the ellipsoid.

If the point is outside the ellipsoid, the lower limit λ in equation (9.4) is the larger root of

$$\frac{x^2+y^2}{a^2+\lambda} + \frac{z^2}{c^2+\lambda} = 1. \tag{9.10}$$

If we write $d^2 = x^2 + y^2$ then this ellipse has its foci on the d axis at $d = \pm ae$ where e, the eccentricity of the original ellipsoid, is given by $e^2 = 1 - c^2/a^2$.

Because the major axis of the ellipse in equation (9.10) is $2\sqrt{a^2 + \lambda}$, and by definition this is the sum of the distances to the two foci from any point on the ellipse, we have

$$2\sqrt{a^2 + \lambda} = \sqrt{z^2 + (d - ae)^2} + \sqrt{z^2 + (d + ae)^2}. \quad (9.11)$$

Since $d = \sqrt{x^2 + y^2}$, λ is a moderately complicated function of x, y, and z. It is straightforward, but tedious, to prove from equation (9.4) with lower limit λ defined by equation (9.10) that $\nabla^2\Phi = 0$. It is easier to obtain the necessary derivatives of λ with respect to x, y, and z by implicit differentiation of equation (9.10) than by explicit differentiation of equation (9.11). This demonstration is left as a problem for enterprising readers.

In terms of the semi-major axis a and the eccentricity e of the original ellipsoid, the integrals in equation (9.4) with $\lambda = 0$ are

$$J_0 = I_0(\lambda = 0) = \frac{2}{ae}\sin^{-1}e, \quad (9.12)$$

$$J_1 = I_1(\lambda = 0) = \frac{1}{a^3 e^2}\left[\frac{1}{e}\sin^{-1}e - \sqrt{1 - e^2}\right], \quad (9.13)$$

and

$$J_2 = I_2(\lambda = 0) = \frac{2}{a^3 e^2}\left[\frac{1}{\sqrt{1 - e^2}} - \frac{1}{e}\sin^{-1}e\right]. \quad (9.14)$$

The potential inside the Maclaurin ellipsoid is

$$\Phi_g(x, y, z) = \pi a^2 cG\rho\left[J_0 - (x^2 + y^2)J_1 - z^2 J_2\right]. \quad (9.15)$$

The potential surfaces inside the original ellipsoid are somewhat less eccentric than the original ellipsoid. Thus the potential surface through the pole at x, $y = 0$, $z = c$ falls short of the equatorial radius a. Therefore the addition of a centrifugal potential due to the systems rotation with angular velocity Ω provides an additional flattening of the equipotential surface, so that the total potential can be constant on the surface of the original ellipsoid. The centrifugal potential is $\Omega^2(x^2 + y^2)/2$. Adding this to the gravitational potential and equating the polar and equatorial values gives

$$\frac{\Omega^2}{\pi G \rho} = a^3 \sqrt{1 - e^2}[J_1 - (1 - e^2)J_2]. \qquad (9.16)$$

This constitutes a fitting of an equipotential to the surface of an ellipsoidal gravitating body of uniform density.

Substituting for J_1 and J_2 gives

$$\frac{\Omega^2}{2\pi G \rho} = \frac{\sqrt{1 - e^2}}{e^3}(3 - 2e^2)\sin^{-1}e - \frac{3}{e^2}(1 - e^2), \quad (9.17)$$

which is Maclaurin's formula. In the limit of small e this is

$$\frac{\Omega^2}{2\pi G \rho} = \frac{4e^2}{15}. \qquad (9.18)$$

For a sphere of radius a and density ρ we have

$$g = 4/3 \pi \rho G a$$

so this becomes $e^2 = 5a\Omega^2/2g$. If we write this in terms of ellipticity $\varepsilon = (a - c)/a$ instead of eccentricity

$$e^2 = (a^2 - c^2)/a^2,$$

we find $\varepsilon = e^2/2$ in the limit of small e. So

$$\varepsilon = \frac{5}{4}\frac{\Omega^2 a}{g} \qquad (9.19)$$

which is equivalent to a 26 kilometer bulge. So you can see what allowing for the mutual gravitational attraction of the bulge can do. Actually we have now got too large a bulge. The real bulge of 21 kilometers can be accounted for by remembering that density is not uniform in the real earth, and actually increases with depth, so that the appropriate value should lie between the 11 kilometer bulge of chapter 7 and the 26 km bulge of Maclaurin's ellipsoid of uniform density. By this time, however, you must be beginning to suspect that the whole quantitative nature of the exact figure of the earth is something of a red herring.*

9.5 PARTICLE MOTIONS ON THE MACLAURIN ELLIPSOID

Let us now consider the problem of computing particle motions on a Maclaurin ellipsoid. This discussion should

*The most modern reference that gives both history and mathematical theory of rotating masses of liquid (including density as function of radius) is Jean-Louis Tassoul, *Theory of Rotating Stars* (Princeton University Press, 1978), but it is at a level of sophistication well beyond the elementary level of this little book of exercises. Another modern treatment is S. Chandrasekhar, *Ellipsoidal Figures of Equilibrium* (New Haven: Yale University Press, 1963; rpt. New York: Dover, 1987). An older reference (nineteenth century) is the second volume of William Thompson and Peter G. Tait, *Treatise on Natural Philosophy* (rpt. New York: Dover, 1962), but we don't recommend it except for the determined scholar. Even the authors succumbed at the prospect of preparing a third volume. We wish that we could present an elementary analytical or economical numerical computer exercise for computing equilibrium figures, but we have not found one.

apply equally well to any oblate ellipsoid as long as the combined effect of gravitational potential and uniform rotation at angular velocity Ω about the z axis make the surface of the ellipsoid an equipotential.

We take the view that the geometry of the ellipsoid and the rate of rotation Ω which make it a potential surface may be taken as parameters, and that we need not take into consideration the relation between Ω and the geometry.

It is convenient to use cylindrical polar coordinates as in chapter 5, and for the time being we adhere to an inertial frame (absolute rest).

Let the equation for the ellipsoid be

$$r^2 = H(z) = a^2 - \frac{z^2}{(1 - e^2)}, \qquad (9.20)$$

where e is the eccentricity of the ellipse. This is equivalent to

$$\frac{r^2}{a^2} + \frac{z^2}{c^2} = 1, \qquad (9.21)$$

with $c^2 = a^2(1 - e^2)$.

We need the gravitational potential on the surface of the ellipsoid, and this leads to the problem of deciding upon a sign convention for potentials. If the force acting on a particle is derivable from a potential Φ, the force is $\bar{F} = \nabla\Phi$. With this sign convention the potential energy is $-\Phi$. To fix this idea consider a mass m under the influence of a uniform gravitational acceleration g acting in the negative z direction. Then F_x and $F_y = 0$, $F_z = -mg$ and $\Phi_g = -mg(z - z_0)$, where $z = z_0$ is an arbitrary reference level. The potential energy in this case is $mg(z - z_0)$, which is the negative of the potential. The kinetic energy of the particle of mass m, described in our cylindrical co-ordinates is $m(\dot{z}^2 + \dot{r}^2 + r^2\dot{\phi}^2)/2$. The total energy, kinetic plus potential, is

$$m(\dot{z}^2 + \dot{r}^2 + r^2\dot{\phi}^2)/2 - \Phi = m[(\dot{z}^2 + \dot{r}^2 + r^2\dot{\phi}^2)/2 + g(z - z_0)]$$

for the case just considered. Compare this with equation (5.2).

On the surface of the ellipsoid we must have

$$\Phi_g + \Phi_c = \text{constant,}$$

where $\Phi_c = m\Omega^2 r^2/2$ is the centrifugal potential for a particle of mass rotating at angular velocity Ω. Therefore the potential energy $-\Phi_g = \Phi_c + \text{const}$ and the total energy conservation for a particle on the ellipsoidal surface may be written as

$$\frac{d}{dt}\left[\frac{m}{2}(\dot{z}^2 + \dot{r}^2 + r^2\dot{\phi}^2 + \Omega^2 r^2)\right] = 0. \qquad (9.22)$$

The other conserved quantity is angular momentum

$$mr^2\dot{\phi} = m\mu.$$

These two equations plus the equation for the surface define the particle's motion. If we denote the velocity component toward the North Pole (on the ellipsoidal surface) as v, then $v^2 = \dot{z}^2 + \dot{r}^2$. Also $u = r\dot{\phi}$ is the eastward component. The energy equation is

$$v^2 + \frac{\mu^2}{r^2} + \Omega^2 r^2 = 2E/m \qquad (9.23)$$

where we have used $r^2\dot{\phi} = \mu$ to rewrite the u^2 term. The initial value of u, v and r determine E and μ. Clearly as r^2 either increases or decreases the left hand side of the equation will become large enough so that

$$v^2 = 2E/m - \frac{\mu^2}{r^2} - \Omega^2 r^2$$

must go to zero. We can characterize the limiting circles by simply giving the value of r and μ corresponding to one of the limiting circles. Let r_1 be one such value, which makes $v^2 = 0$. The other root of the right hand side of the v^2 equation is then $r_2^2 = \mu^2/(\Omega^2 r_1^2)$, since the product of the roots for r^2 must be μ^2/Ω^2.

To see what this means, call the angular velocity (at $r = r_1$) $\dot{\phi} = \omega$. Then $\mu = \omega r_1^2$ and $r_2^2 = (\omega^2/\Omega^2)r_1^2$. If $\omega^2 < \Omega^2$ the limiting circle $r = r_1$ is largest radius the particle reaches and it oscillates between r_1 and

$$r_2 = |\omega/\Omega| r_1 .$$

At $r = r_2$ the angular velocity is $\dot{\phi} = \Omega^2/\omega$ to conserve angular momentum. If $\omega^2 > \Omega^2$, r_2 is larger than r_1. If $|\omega r_1| > \Omega a$ then r_2 will be greater than a. The particle then crosses the equator and comes back to the radius r_1 in the other hemisphere. Note the condition for this to happen. The linear velocity $u = \omega r_1$ at the limiting circle must be larger than the speed of an equatorial expert particle [Ωa].

PROBLEM

Problem 9.0 The acceleration of gravitation alone within the interior of a rotating sphere of radius a is gr/a where

g is the value at $r = a$. Bore a hole to the center of the earth from the pole, and another from the equator, that meets it at the center of the earth. Slowly fill the hole at the equator with water until the water level reaches the surface. Compute the level in the hole at the pole, and show that it lies beneath the surface by an amount $\Omega^2 a^2 / 2g$.

EXERCISES

Exercise 9-1 [lines 47000–47220] *Three kinds of potential.* This program draws, by color contouring, three potentials in the x, z plane. They really are surfaces of revolution, but we look only at a cut through a plane. The vertical z axis is the axis of rotation. When $K = 1$ [47020] the program sets $A = A0$ and $B = 0$ [47030]. This causes the potential [47100] to correspond to a purely centrifugal, or axifugal one, with amplitude $A0$. Depending upon the amplitude of the potential a color is chosen [47110] and a little square with this color is plotted [47120–47150] in each of the symmetrical quadrants of the x, z plane. This process is located in a loop [47060–47170] so that the whole plane is covered within limits X, Y -2, -2 to 2, 2. At the end of the plot the program goes to a subroutine [47200–47220] that cleans up the center and provides a delay [47210] to allow for time to view the picture. Then the program goes back to [47020] takes the next K and then proceeds to draw up the picture of the central gravitation potential alone. During the next cycle, $K = 3$, both potentials are added, so this time we see a figure of the combined potential, corresponding to figure 9.1.

```
47000 'Exercise 9-1:  Equipotential surfaces
47005 SCREEN 1: COLOR 0,2: KEY OFF: CLS
47010 A0= 2: B0= 10
47020 FOR K = 1 TO 3:CLS:LOCATE 5,28
47030 IF K = 1 THEN A=A0:B=0:PRINT"axifugal"
47040 IF K= 2 THEN A = 0: B = B0
47041 IF K= 2 THEN PRINT "gravitational"
47050 IF K = 3 THEN A = A0:B=B0
47051 IF K = 3 THEN PRINT"combined"
47060 FOR X =0   TO 2   STEP .05
47070 FOR Y = 0   TO 2   STEP .05
47080 R = SQR(X^2+Y^2)
47090 IF R<.6 THEN GOTO 47160
47100 POT = A *X^2 + B/R
47110 COLEUR = INT(POT) MOD 3 +1
47115 I=50*X:J=50*Y:CO=COLEUR
47120 LINE(100+I,100-J)-(102+I,102-J),CO,BF
47130 LINE(100-I,100-J)-(102-I,102-J),CO,BF
47140 LINE(100-I,100+J)-(102-I,102+J),CO,BF
47150 LINE(100+I,100+J)-(102+I,102+J),CO,BF
47160 NEXT
47170 NEXT  :GOSUB 47200
47180 NEXT
47190 GOTO 47010
47200 CIRCLE (101,101),34,0:PAINT( 70,101),0
47210 FOR Z = 1 TO 2000: ZZ=SQR(Z):NEXT
47220 RETURN
```

Exercise 9-2 [lines 48000–48390] *Particle on Maclaurin ellipsoid.*

The equation used for the computation in absolute inertial space is the differentiated form of equation (9.22). This is more convenient computationally than use of the undifferentiated form because it is not necessary to cope with choices of sign of the square root of $\dot{z}^2$. The numerical integration of $\ddot{z}$ is in lines 48160–48240. The equation integrated is written in the form $c_1\ddot{z} = c_2 + c_3$ where

$$c_1 = -\frac{c^2 r^2}{a^2 z}\left(1 + \frac{a^4 z^2}{c^4 r^2}\right),$$

$$c_2 = \dot{z}^2\frac{a^4}{c^4}\left(\frac{c^2}{a^2} + \frac{z^2}{r^2}\right),$$

$$c_3 = \frac{\mu^2}{r^2} - \Omega^2 r^2.$$

The space coordinates x and y are computed [48260–70], and then translated into plotting coordinates L, M (absolute) and LR, MR (relative to the rotating ellipsoid) [48280–90]. A marker fixed to the equator, XM, YM and its plotting coordinates LM, MM are also computed [48300–48310] for visualizing easily how fast the ellipsoid is rotating in inertial space. The plotting coordinates show a distant view of the ellipsoid from a point at infinity and the angle INC [48050] above the equatorial plane. You may want to change INC in certain cases. The *SLOPE* of the surface of the ellipsoid dz/dr is computed [48340] at the instantaneous position of the particle so that we can get the geographic latitude $LATG$. The geocentric latitude LAT [48350] is also computed and values printed on screen [48350–60]. The points are plotted [48320–30, 48380] and the loop returned [48390] to 48160 for the next time step of integration. Parameters may be reassigned [48010–48050]. The visual outline of the ellipsoid is computed [48060–48100] and plotted [48120], and also the equator [48110], as is the tilted line segment of the polar axis inside the ellipsoid [48130–40].

```
48000 ' 9-2 Maclaurin ellipsoid *************
48001 ' in cylindrical coordinates
48010 DT = .02: A = 2: C = 1  : Z = .8*C
48011 OMEGA = 1: PRD=.3 :ZD=0
48020 R2=(1-Z^2/C^2)*A^2 : R = SQR(R2)
48030 MU = (PRD+OMEGA)*R2
48040 SCREEN 1: COLOR 0,2: KEY OFF: CLS
48045 WINDOW SCREEN (0,0)-(640,400)
48050 PI = 3.14159: INC = PI/10
48060 Q = 1+(TAN(INC)*A/C)^2
48070 ZO = C/SQR(Q)
48080 RO = (A/C)^2*ZO*TAN(INC)
48090 B = ZO+RO*TAN(INC)
48100 D = B*COS(INC)
48110 CIRCLE(300,200),100*A,1,,,SIN(INC)
48120 CIRCLE(300,200),100*A,1,,,D/A
48130 LINE(300,200)-(300,200+100*C*COS(INC)),1
48140 LINE(300,200)-(300,200-100*C*COS(INC)),1
48150 GOTO 48260
48160 R2=(1-Z^2/C^2)*A^2 : R = SQR(R2)
48170 PD = MU/R2
48180 P = PD*DT+P : PR = P -T*OMEGA
48190 AAA=(1+(A^4*Z^2)/(C^4*R2))
48191 C1= -(C^2/A^2)*(R2/Z)*AAA
48200 C2=(ZD^2*A^4/C^4)*((C^2/A^2)+(Z^2/R2))
48210 C3= (MU^2/R2)-(OMEGA^2*R2)
48220 ZDD=(C2+C3)/C1
48230 ZD=ZDD*DT+ZD
48240 Z = ZD*DT+Z
48250 T = T +DT
48260 X = R*SIN(P):Y=-R*COS(P)
48270 XR = R*SIN(PR): YR = -R*COS (PR)
48280 L = 100*X+300
48281 M=-100*(Y*SIN(INC)+Z*COS(INC))+200
48290 LR= 100*XR+300
```

```
48291 MR=-100*(YR*SIN(INC)+Z*COS(INC))+200
48300 XM=A*SIN(T):YM=-A*COS(T)
48310 LM=100*XM+300
48311 MM=-100*(YM*SIN(INC))+200
48320 PSET(L,M),3
48330 PSET(LR,MR),2
48340 SLOPE = ATN(-R*C^2/(Z*A^2))
48341 LATG= SLOPE+PI/2
48350 LAT = ATN(Z/R)*360/(2*PI)
48360 LOCATE 2,1
48361 TT=T*OMEGA/(2*PI)
48362 LG=LATG*360/(2*PI)
48363 LN=P*360/(2*PI)
48364 LR=PR*360/(2*PI)
48370 PRINT USING "t=##.## lat=###.#";TT,LAT
48371 LOCATE 2,21
48372 PRINT USING "latg=##.# lon=###.#";LG,LN
48373 LOCATE 3, 1
48374 PRINT USING "lonr=###.#";LR
48380 PSET(LM,MM),3
48390 GOTO 48160
```

Exercise 9-3 [lines 49000−49510] *Gravitational potential of the Maclaurin ellipsoid.* The size of the major and minor semiaxes of the ellipsoid whose gravitational potential surfaces are to be displayed in the *x, z* plane (we set *y* = 0) are set in line 49020 and the outline of the ellipse drawn [49030]. The lambda variable is called *L* in the program, and is initially zero, so when we compute the dummy constants *U* and *V* [49040−50], the values of the integrals *J0, J1, J2* (equations 9.12−9.14) [49060−80]. These are stored for later computation of the potential *PHI* inside the Maclaurin ellipsoid.

The potential outside the ellipsoid is computed by choosing discrete steps in *PHI* [49160] for contours of equipotential curves in the *x, z* plane, and stepping through lambda, *L*, [49170], computing the integrals *I0, I1, I2* (equations 9.5, 9.7 and 9.8) [49180−49220], and then finding *z* from

$$z^2 = \frac{\Phi - I_0 + (a^2 + \lambda)I_1}{I_1[(a^2 + \lambda)/(c^2 + \lambda)] - I_2}.$$

This form [49230] is obtained by substituting *x* from

$$x^2 = (a^2 + \lambda)\left(1 - \frac{z^2}{c^2 + \lambda}\right),$$

where *y* = 0 into equation (9.4′). We can then compute *x* [49270−90] and plot the contour for the desired value of *PHI* [49300−30]. The loop is repeated for all *L* and *PHI* steps [49360−70].

We then proceed to compute the equipotential contours inside the Maclaurin spheroid [49380−500] in a similar loop, but in this case with *L* = 0 so we use *J0, J1,*

$J2$ instead of $I0$, $I1$, $I2$. The procedure is similar except that now we step in Z, and the only difference is the necessity to cut off the loop when the point computed lies outside the Maclaurin ellipsoid. This cutting off is done by the *IF* intercept on W [49430].

```
49000 ' Exercise 9-3  Gravitational potential
49001 ' of Maclaurin ellipsoid
49010 SCREEN 1:COLOR 0,2:KEY OFF:CLS
49015 WINDOW SCREEN (0,0)-(640,400)
49020 A = 2  :C=1  : A2 = A^2: C2=C^2 :L=0
49030 CIRCLE (300,200),100*A,1,,,C/A
49035 LOCATE 1,30:PRINT "WAIT"
49040  U = SQR((A2-C2)/(A2+L))
49050  V =ATN(U/SQR(-U*U+1))
49060  J0 = 2*V/SQR(A2-C2)
49065  AAA=((A2-C2)*SQR(C2+L))
49066  AAA=-1/AAA+J0/(2*(A2-C2))
49070  J1=1/((A2+L)*SQR(C2+L))+AAA
49080  J2= (2/SQR(C2+L)-J0)/(A2-C2)
49090  FOR Z = 0 TO 1  STEP .099
49100   X = A*SQR(1-Z^2/C2)
49110   PHI = I0-X^2*I1-Z^2*I2
49140  NEXT
49160 FOR PHI=.9  TO  .5 STEP -.05
49170 FOR L= .01 TO 3 STEP .02
49180  U = SQR((A2-C2)/(A2+L))
49190  V =ATN(U/SQR(-U*U+1))
49200  I0 = 2*V/SQR(A2-C2)
49205  AAA=((A2-C2)*SQR(C2+L))
49206  AAA=-1/AAA+I0/(2*(A2-C2))
49210  I1=1/((A2+L)*SQR(C2+L))+AAA
49220  I2= (2/SQR(C2+L)-I0)/(A2-C2)
49225  BBB=(I1*(A2+L)/(C2+L)-I2)
49230 Z2 = (PHI-I0+(A2+L)*I1)/BBB
49250 IF Z2<0 THEN GOTO 49360
49260 Z = SQR(Z2)
49270 X2=(A2+L)*(1-Z2/(C2+L))
49280 IF X2<0 THEN GOTO 49360
49290 X = SQR(X2)
49300 PSET(300-100*X,200-100*Z),7
```

```
49310 PSET(300+100*X,200-100*Z),7
49320 PSET(300-100*X,200+100*Z),7
49330 PSET(300+100*X,200+100*Z),7
49360 NEXT L
49370 NEXT PHI
49380 FOR PHI = 1.2 TO .6 STEP -.05
49390 FOR Z = 0 TO 1 STEP .02
49400 X2=(J0-Z^2*J2-PHI)/J1
49410 IF X2 <0 THEN GOTO 49500
49420 X = SQR(X2)
49430 W = X2/A2+Z^2/C2
49440 IF W>1 THEN GOTO 49490
49450 PSET (300-100*X,200-100*Z),7
49460 PSET (300-100*X,200+100*Z),7
49470 PSET (300+100*X,200+100*Z),7
49480 PSET (300+100*X,200-100*Z),7
49490 NEXT Z
49500 NEXT PHI
49510 STOP
```

Exercise 9-4 [54000–54180] *Rotation rate of Maclaurin ellip-soid.* This program is in double precision [54010]. We are invited to enter a value of the eccentricity [54040] be-tween 0 and 1. We then use the equations (9.17) [54050–54140] to compute the equilibrium rate of rotation, time of revolution for a mass equivalent to the of the earth, and ellipticity, printing these values on the screen [54150–60]. substituting LPRINT for PRINT will print on type-writer. Then [54180] returns us to [54040] for another input of *e*. You will discover that there is a maximum ro-tation for an intermediate value of *e*. The values computed correspond to those given in the table on page 703 of Lamb's *Hydrodynamics*.

```
54000 'Exercise 9-4: rotation rate******
54001 'of Maclaurin ellipsoid
54010 DEFDBL A-D,F-Z
54020 PI = 4*ATN(1)
54030 SCREEN 2:KEY OFF:CLS
54040 PRINT "enter e";:INPUT E
54050 A = (1/(1-E^2))^(1/6)
54060 C = A*SQR(1-E^2)
54070 A2=A^2:C2=C^2
54080  V =ATN(E/SQR(-E*E+1))
54085 AAA=3*(1-E^2)/E^2
54090 ZZ=SQR(1-E^2)*(3-2*E^2)*V/(E^3)-AAA
54100 OMEGA2 = ZZ*2*PI*5.53*6.67E-08
54110 OMEGA = SQR(OMEGA2)
54120 T=2*PI/OMEGA
54130 HRS = T/3600
54140 IHRS=INT(HRS):MIN=60*(HRS-IHRS)
54150 PRINT USING"e=#.### a=#.### c=#.###";E,A,C
54155 PRINT USING"omega2/2pi*rho= #.#####";ZZ
54160 PRINT USING"period=###h ##m";IHRS,MIN
54165 PRINT USING"ellipticity=1/###.##";A/(A-C)
54170 CIRCLE(500,160),50*A,7,,,C/A*.5
54180  GOTO 54040
```

CHAPTER X

Concluding remarks

10.1 GENERAL REFERENCES—
OTHER PLACES TO LOOK

For students who are fluent in vector analysis the idea of Coriolis force may present no sense of mystery. We have not searched the many present day textbooks of physics, engineering mechanics, and geophysics that treat Coriolis force, but can point to two recent texts that contain succinct derivations of expressions for acceleration in rotating reference frames, use them to form equations of motion in hydrodynamical form, and introduce spherical coordinates for application to problems of atmospheric and oceanic motion: Adrian Gill, *Atmosphere—Ocean Dynamics* (New York: Academic Press, 1982), pp. 72–76 and Joseph Pedlosky, *Geophysical Fluid Dynamics* (New York: Springer-Verlag, 1979), pp. 14–20. The derivations are similar to ones used in texts for general theoretical physics, for example *Theoretical Physics* by Georg Joos, (English version, New York: G. E. Stechert & Co., 1934, pp. 219–223), but are more explicit about how the centrifugal force due to the rotation of the earth may be absorbed as a small term into the potential of Newtonian gravitation to form a new potential that yields what we commonly call gravity. The dynamics of rotating systems is covered in Horace Lamb, *Hydrodynamics* (Cambridge: Cambridge University Press, 6th ed., 1932), pp. 307ff, 330ff, and the important chapter on the figures of equilibria of rotating masses of liquid (pp. 697–730).

General textbooks on oceanography of a former generation, such as H. U. Sverdrup, M. Johnson, and R. Fleming, *The Oceans* (Englewood Cliffs, N.J.: Prentice-Hall); Albert Defant, *Physical Oceanography* 2 (New York: Pergamon Press, 2 vols, 1961) finesse an explanation of Coriolis force by referring the student to V. Bjerknes et al., *Physikalische Hydrodynamik* (Berlin: Springer, 1933). Those who rush to the library for enlightenment will find themselves confronted by a formidable treatise in the German language in which rotating systems first arise in chapter 12. We do not think that referring students to such an immense treatise at a crucial point, without detailed guidance, is a very funny joke. Other texts on oceanography give no specific reference at all.

Meteorological texts are more professional, and there is no shortage of good exposition in them. One of the most complete modern discussions is by Norman A. Phillips in the first chapter of P. Morel, ed., *Dynamical Meteorology* (Amsterdam: Reidel, 1973).

Because our examples deal entirely with trajectories of particles on various platforms and not with continua, we are wandering over well-trodden ground. Probably every example we produce can be found somewhere in the nineteenth-century literature. A good study of particle motions on the rotating globe was published by F. J. W. Whipple in *Philosophical Magazine*, 1917, pp. 457ff. So we make no claim to scientific originality.

10.2 A VECTOR DERIVATION

The standard derivation of the acceleration in a uniformly rotating system of coordinates uses some vector analysis. Suppose that we use unprimed quantities to indicate variables referred to the absolute inertial reference system, and primed quantities for variables defined in terms of a system rotating uniformly with vector spin Ω. A point P with constant position vector $\mathbf{r}'$ in the rotating system thus has an absolute velocity $\Omega \times \mathbf{r}'$. When the point also has a velocity $d\mathbf{r}'/dt$ relative to the rotating frame, then the velocity in the resting system is the sum of the two, in the identity (true, in fact, for any vector).

$$\frac{d}{dt}\mathbf{r} \equiv \frac{d}{dt}\mathbf{r}' + \Omega \times \mathbf{r}'. \qquad (10.1)$$

If we repeat this vector operation again we obtain the acceleration in absolute resting space on the left hand side of the identity and on the right we find the various terms of which it is composed, written in terms of variables measurable with respect to the moving frame.

$$\frac{d^2}{dt^2}\mathbf{r} \equiv \frac{d}{dt}\left(\frac{d}{dt}\mathbf{r}' + \Omega \times \mathbf{r}'\right)$$
$$+ \Omega \times \left(\frac{d}{dt}\mathbf{r}' + \Omega \times \mathbf{r}'\right), \qquad (10.2)$$

or

$$\frac{d^2}{dt^2}\mathbf{r} \equiv \frac{d^2}{dt^2}\mathbf{r}' + 2\Omega \times \frac{d\mathbf{r}'}{dt} + \Omega \times (\Omega \times r'). \qquad (10.3)$$

Here again, on the right hand side of the identity, we encounter one of those complicated expressions for acceleration that we have become accustomed to. The above expression says that the acceleration in the fixed system is expressible as the sum of three terms. The first is the conventional expression for an acceleration, written with respect to the moving system (but as though we did not know it is moving). The second term depends upon the velocity relative to the rotating system, and is called the Coriolis acceleration. The third term is the centripetal acceleration. (It would be better to call it an axipetal acceleration because it represents, as we will see, acceleration toward the Ω axis and not toward the origin.)

Both sides of the identities are accelerations. When we write down a dynamical equation for the motion of a particle under a real force (per unit mass) **F**, we have in absolute inertial space

$$\frac{d^2}{dt^2}\mathbf{r} = \mathbf{F}. \qquad (10.4)$$

If the force **F** is more easily specified as a force **F′** in the uniformly rotating reference frame we have used the expanded form of the acceleration on its left hand side [this is of course the right hand side of the identity (10.3)].

$$\frac{d^2}{dt^2}\mathbf{r}' + 2\mathbf{\Omega} \times \frac{d\mathbf{r}'}{dt} + \mathbf{\Omega} \times (\mathbf{\Omega} \times \mathbf{r}') = \mathbf{F}'. \qquad (10.5)$$

If we move the two terms for acceleration that involve Ω to the right hand side of this dynamical equation we have

$$\frac{d^2}{dt^2}\mathbf{r}' = -2\mathbf{\Omega} \times \frac{d\mathbf{r}'}{dt} - \mathbf{\Omega} \times (\mathbf{\Omega} \times \mathbf{r}') + \mathbf{F}', \qquad (10.6)$$

where the first term on the right hand is now called a Coriolis force due to the rotation $\mathbf{\Omega}$ and the relative velocity $(d/dt)\mathbf{r}'$, and the second term is called the centrifugal force due to the rotation of the system. This gives the left hand side of the equation the form of a Newtonian dynamical system in inertial space, but at the expense of introducing two virtual (or apparent) forces. Because $\mathbf{F}'$, as we presupposed above, is most conveniently defined in terms of the rotating reference frame the formulation involving virtual forces is often more economical analytically than that in inertial space. For geophysical phenomena on the rotating earth it has considerable conceptual advantages.

Meteorologists and oceanographers study small motions in thin layers of gas or liquid on the surface of a rotating earth.

10.3 SIZE OF ACCELERATIONS AND FORCES IN TERRESTRIAL FLUIDS

We now consider fluid layers on the rotating earth. We will estimate the size of the various terms in the equations and show that the Coriolis forces are important. We have to introduce the idea of pressure, with a vertical pressure gradient acting to counterbalance "gravity." When one considers how shallow the atmosphere and ocean are on the earth, then range of altitude involved is so small that the equipotential surfaces are nearly parallel.

The equations of motion in the rotating spherical coordinate system (6.25–6.27) with the forces $F'_E = 0$, $F'_N = R \sin \vartheta \cos \vartheta\ \Omega^2$ and $F'_V = g' = g + R \cos^2\vartheta\ \Omega^2$ added on the right hand side to produce a rotating equilibrium at angular velocity Ω permit the relative velocities

u', v' and w' all to vanish, so that the particles are at relative rest. The equations can now be augmented with pressure gradients on the right hand side, so that they now read

$$\dot{u}' - \frac{u'v'}{R}\tan\vartheta + \frac{u'w'}{R} = \frac{-\alpha}{R\cos\vartheta}\frac{\partial p}{\partial\phi} - 2\Omega\cos\vartheta\, w'$$
$$+ 2\Omega\sin\vartheta\, v'$$

$$\dot{v}' + \frac{u'^2}{R}\tan\vartheta + \frac{v'w'}{R} = -\frac{\alpha}{R}\frac{\partial p}{\partial\vartheta} - 2\Omega\sin\vartheta\, u'$$

$$\dot{w}' - \frac{u'^2 + v'^2}{R} = -\alpha\frac{\partial p}{\partial R} + 2\Omega\cos\vartheta\, u' - g' \quad (10.7)$$

where α is the specific volume of the fluid.

In terms of these equations we can estimate the magnitudes of the various terms in the accelerations relative to the earth that are experienced in studying the motions of the air and ocean. We choose values of wind velocity, horizontal scale, depth, and period to correspond to a synoptic feature in the air—a moving weather system. In the ocean we choose two cases. The first is one of those eddies that surround strong currents like the Gulf Stream —a so-called mesoscale eddy. The second ocean feature will be one of the large circulation gyres, nearly stationary in time.

In table 10.1 we list representative values of the horizontal velocity U, the vertical velocity W, a length-scale L (horizontal wavelength/2π), a depth scale D, a time scale T (period/2π). At mid-latitude we can choose both components of the Coriolis parameter to be the same $2\Omega\sin\vartheta \cong 2\Omega\cos\vartheta \cong 10^{-4}\,\text{sec}^{-1}$ and the radius of the earth a as $6\times10^6\,m$. Everything will be in meters and seconds. We choose these magnitudes for each of the

TABLE 10.1. Representative magnitudes of variables

	ATMOSPHERE	OCEAN		UNITS
		EDDY	GYRE	
U	10	10^{-1}	10^{-2}	m s^{-1}
W	10^{-2}	3×10^{-6}	3×10^{-7}	m s^{-1}
L	10^6	10^5	10^6	m
D	10^4	10^4	10^3	m
T	10^5	10^6	∞	s

TABLE 10.2. Vertical Accelerations, m s^{-2}

	ATMOSPHERE	OCEAN	
		EDDY	GYRE
dw'/dt	10^{-7}	3×10^{-11}	0
$2\Omega \cos \vartheta \, u'$	10^{-3}	10^{-5}	10^{-6}
$(u'^2 + v'^2)/R$	10^{-5}	10^{-9}	10^{-11}

three cases: for the air a synoptic weather disturbance, for the ocean a mesoscale eddy, and a subtropical ocean-wide gyre of the major current systems.

Using the equations (10.7) to identify various terms of the acceleration in the terrestrial spherical coordinate format, we proceed to Table 10.2, which gives the magnitudes of the various terms in the vertical acceleration.

TABLE 10.3A. Eastward acceleration terms, m sec^{-2}

	ATMOSPHERE	OCEAN	
		EDDY	GYRE
du'/dt	10^{-4}	10^{-7}	0
$2\Omega \sin \vartheta\, v'$	10^{-3}	10^{-5}	10^{-6}
$2\Omega \cos \vartheta\, w'$	10^{-6}	3×10^{-10}	3×10^{-11}
$u'w'/R$	10^{-8}	3×10^{-14}	3×10^{-16}
$u'v' \tan \vartheta/R$	10^{-5}	10^{-9}	10^{-11}

TABLE 10.3B. Northward acceleration terms, m sec^{-2}

	ATMOSPHERE	OCEAN	
		EDDY	GYRE
dv'/dt	10^{-4}	10^{-7}	0
$2\Omega \sin \vartheta\, u'$	10^{-3}	10^{-5}	10^{-6}
$v'w'/R$	10^{-8}	3×10^{-14}	3×10^{-16}
$u'^{2} \tan \vartheta/R$	10^{-5}	10^{-9}	10^{-11}

These terms are all quite small compared to gravity of the order $g' = 10$ ms^{-2}. Obviously in both atmosphere and ocean none of the acceleration terms in table 10.2 come anywhere near to gravity, so the vertical presure gradient must balance it. Because the other terms are so very small the vertical balance in the air and ocean for phenomena of these kinds, must be essentially hydrostatic.

We turn to an estimate of the terms in the horizontal accelerations, as displayed in table 10.3.

10.4 PRESSURE GRADIENTS

As can be seen the largest accelerations in all three cases and in both eastward and northward components are the Coriolis accelerations due to horizontal motions. In the weather system of the air the local time derivative is not entirely negligible, being only one order of magnitude less than the Coriolis acceleration. In both oceanic cases the Coriolis accelerations due to horizontal motion far exceed any other acceleration. They need to be balanced by some force in the full dynamical equations of course, and the main contender is always the horizontal pressure gradient, due to the fluids' surface and internal density structure not being strictly parallel to gravity potential surfaces. It is the balance of the Coriolis acceleration and the horizontal pressure gradient that characterizes most of the phenomenology of winds and currents and their relation to regions of high and low pressure.

In order to get some idea of the magnitude of the pressure variation ΔP from one point to another that provides the force needed to balance the dominant terms in the accelerations, let us first consider the vertical balance. Because the force of gravity per unit mass is $g' = 10 \text{ ms}^{-2}$, it is evident that none of the terms in the vertical acceleration of table 10.1 are big enough, so we consider the vertical difference of pressure, ΔP over a depth H (let us say 10 meters). To make the pressure gradient $\Delta P/H$ into a force per unit mass we must multiply it by the specific volume α of the air or ocean water. For order of magnitude purposes, near the earth's surface

$$\alpha_{\text{atmosphere}} \cong 1 \, \text{m}^3\text{kg}^{-1}, \qquad \alpha_{\text{water}} = 10^{-3}\text{m}^3\text{kg}^{-1}.$$

The units are the standard *SI* units. If we choose first water, then the increase of pressure ΔP with altitude H is

$$\Delta P_{\text{water}} = -10^5 \text{Pascals}.$$

The old fashioned unit of pressure, the bar, is equivalent to 10^5 Pascals, so we see, as we always knew, that on the surface of the earth, the height of a column of water 10 meters high has a pressure on the bottom of one bar —or one "atmosphere"—or even more old-fashionedly, 15 pounds per square inch. The rate at which the air pressure decreases with altitude at sea level is much less because of the greater specific volume. Assuming that the air is not compressible, and that it has the same density as we go up, we would reach the top of the air at 10,000 meters. If the density of ocean water were uniform with depth the pressure would increase by one bar for every ten meters as we descended. Meteorologists prefer the millibar (one one thousandth of a bar) as a pressure unit. Oceanographers fancy the decibar, because it makes pressure and depth (in meters) nearly equal in the ocean.

When we come to the horizontal accelerations we will make use of horizontal pressure gradients: pressure differences of ΔP over horizontal length scales L. From Table 10.3 we already know that the leading terms are those caused by Coriolis accelerations due to horizontal components of the relative velocity U. Denoting the magnitude of these leading terms by A_L we can construct table 10.4 to compute the corresponding ΔP.

In all three cases we find 10^3 Pascals, or 10 millibars. At the ocean surface these horizontal pressure gradients could be produced by deviations of the sea surface from a level surface of only 0.1 meter. Actually, in the case of a

TABLE 10.4. Computation of pressure difference

	ATMOSPHERE	OCEAN		UNITS
		EDDY	GYRE	
A_L	10^{-3}	10^{-5}	10^{-6}	m s^{-2}
L	10^6	10^5	10^6	m
α	1	10^{-3}	10^{-3}	m^3 kg^{-1}
ΔP	10^3	10^3	10^3	Pascals

subtropical gyre the east-west dimensions are 5 to 10 times as large as the north-south ones that were used to define L in the previous tables, so east-west deviations from sea level can be as much as a full meter. If one takes a global survey of both atmosphere and ocean, the horizontal pressure differences in the ocean exceed those in the atmosphere. It is these rather small differences of pressure, balanced by horizontal Coriolis forces, that dominate the dynamics of weather and ocean currents on our earth. Because of the balance the pressure differences tend to change much more slowly than they would if they dispersed as gravity waves.

APPENDIX

The Compton generator

A.1 HISTORICAL BACKGROUND

While still an undergraduate at Wooster (Ohio) College, the celebrated experimental physicist Arthur Holly Compton [*Science* (1913) 37: 803–6] conducted a novel laboratory demonstration to show that the earth rotates. He built a ring of glass tubing with radius one meter, and filled it with water and oil droplets. After the tube lay horizontal until the fluid became as quiescent as possible, he turned the ring abruptly [over an interval of about three seconds] about an east-west axis, and then measured the speed of drift V [about 0.05 millimeters per second] of the oil droplets by micrometer and metronome. Using an argument (which seems uncertain to us) about differential linear momentum conservation in absolute space, he arrived at a formula for an expected value of the velocity of droplets in the tube $V = r\Omega \sin \vartheta$ where r is the radius of the ring, Ω is the angular velocity of the earth, and ϑ the latitude of Wooster. The correct formula seems to us to be $V = 2r\Omega \sin \vartheta$.

Problems in rotating coordinates are notorious for missing factors of two, so unless we have misread something, or Compton made a compensating error, we cannot account for his getting the earth's rotation rate correctly. His determination of the latitude of Wooster is not in question, nor the azimuth. One supposes that the obscurity of the theoretical formulation confused contemporary geophysicists and that is why Compton's wonderful experiment is not mentioned in meteorological or oceanographical texts.

When Compton entered graduate school at the Palmer Physical Laboratory at Princeton he was invited to repeat the experiment with new apparatus, in an insulated box at 4 degrees centigrade, and further refinements. He obtained the earth's rotation rate to 3%, the latitude to 2.4 degrees and the azimuth to 1.4 degrees [*Physical Review* (February 1915) 5 (2): 109–117; *Scientific American Supplement* no. 2047 (March 7, 1915), pp. 196–197].

Our purpose is not to pursue the apparent discrepancy, nor to repeat this difficult and ingenious experiment, but to use Compton's idea as a stimulus for some further thoughts and experiments that are interesting to perform. First we offer our own, rather crude, derivation of the formula for expected initial velocity in the tube, according to the principles involved as we understand them.

A.2 COMPUTATION OF FLOW IN THE COMPTON EXPERIMENT

The formula can be obtained from equations (10.7) for the Coriolis forces. According to figure A.1, as the ring is rotated by lifting its southern rim and moving it northward around the east-west axis, you can expect an eastward Coriolis force on the water in the tube, and in the lower half of the ring, where the movement is to the south, a westward Coriolis force. The velocity of the water in the tube is essentially confined to the plane of the ring,

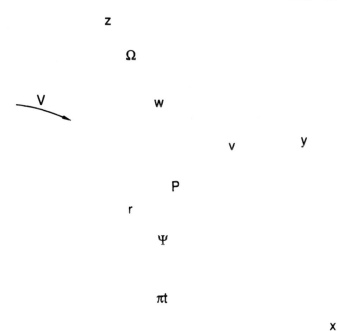

Figure A.1. Coordinate system rotating about the vertical axis, z, with angular velocity Ω. Initially the Compton ring of tubing lies in the horizontal x, y plane. The ring is cranked around the x-axis by the amount πt. At a point P on the ring the velocity components relative to the rotating coordinates are v and w. V is the velocity of fluid along the bore of the tube. The position of the point P in the plane of the ring is defined by r and ψ.

as the induced flow around the tube is very small. So we can neglect the eastward velocity component u and take v and w as given by the motion of the overturning ring. Choose a point P on the tube, at angular distance ψ from the eastern end of the east-west axis about which the tube is turned. The angle through which the tube is turned in the meridional plane (at a uniform rate) is πt so that the turning begins at $t = 0$ and ends at $t = 1$. The y and z coordinates of point P are

$$y = -r\sin\psi\cos\pi t,$$
$$z = r\sin\psi\sin\pi t, \qquad (A.1)$$

and the velocity of the tube at P is

$$v = r\pi\sin\psi\sin\pi t,$$
$$w = r\pi\sin\psi\cos\pi t. \qquad (A.2)$$

With $u = 0$ examine the equations for u, v, w on the spheroidal platform (equation 10.7) and you discover that only in the $\dot{u}$ is there any Coriolis force: in fact there are two

$$-2\Omega\cos\vartheta w + 2\Omega\sin\vartheta v. \qquad (A.3)$$

But we can see that because w reverses sign halfway through the turn, the Coriolis force due to the vertical motion w contributes nothing to the time-integrated eastward force, so effective Coriolis force is simply

$$2\Omega\sin\vartheta r\pi\sin\psi\sin\pi t. \qquad (A.4)$$

The acceleration of fluid in the tube depends upon the component of this force tangential to the tube

$$2\Omega\sin\vartheta\, r^2\pi \int_0^1 dt \int_0^{2\pi} \sin^2\psi\sin\pi t\, d\psi, \qquad (A.5)$$

integrated around the tube (ψ from 0 to 2π) and over the time interval of the overturn ($t = 0$ to 1)

$$2\Omega\sin\vartheta \cdot 2r^2\pi. \qquad (A.6)$$

Dividing this by the circumference of the ring we have the net circumferential impulse on the water in the tube, which of course, because of the incompressibility of the water, all must go at the same circumferential velocity V

$$V = 2\Omega r\sin\vartheta. \qquad (A.7)$$

This result, of course, depends upon a vastly oversimplified abstract picture of the true physical situation. The three-dimensional velocity of a particle of water in the torus is certainly not a simple function of r and t. When the torus begins to turn over, thin frictional boundary layers must form near the walls of the tubing. A closed contour initially drawn through marked particles of water will twist, perhaps fold upon itself in a very complicated way. So far as we know a complete mathematical analysis of this experiment has never been made. So we are not on completely solid logical ground with our own factor of two, even for laminar flow.

A.3 DO IT YOURSELF

You can conduct an interesting qualitative Compton experiment illustrating Coriolis force if you can rotate yourself steadily while standing up without getting too dizzy.

Ballroom dancers take heart. You will need about two feet of flexible transparent plastic hose (Nalgen is good) of about 3/4 inch inside diameter, and a short piece that is slightly bigger in diameter so that it nests tightly on the outside. Form a ring (torus) and join it by slipping both ends into the larger short section. If the tubing is freshly cut from a coil it will bend smoothly into a torus. Put a few tealeaves inside to help visualize the flow, and fill it completely with water by submerging and closing entirely in a sink or pail. This is a piece of equipment that you can experiment with (figure A.2).

Hold the tube in a horizontal plane and then start spinning yourself steadily around a vertical axis. Then when the tea leaves indicate that the water is at rest with respect to your rotating system, and being careful to keep turning yourself around without interruption despite the growing sense of vertigo, quickly turn the torus over 180 degrees into the horizontal plane again. For a few seconds the fluid flows around the torus in the direction opposite to your rotation. It actually seems to rotate twice as fast as you are going around as viewed by the rotating you. You can verify this by observing how long it takes for the leaves to pass two marks on the torus an eighth or a quarter of the way round the circumference. If you try to time the leaves over a longer arc you will be in trouble with the decay of the relative velocity due to frictional dissipation. The result does not differ whether you turn the torus over around a horizontal axis perpendicular or parallel to your belly.

If you start with the torus in a vertical plane, however, and turn it 180 degrees around any vertical axis you will see no motion induced in the torus.

Pierre Flament and one of us (H.S.) tried this experiment with a torus 10 inches in diameter made of 3/4" transparent tubing filled with water and some tealeaves.

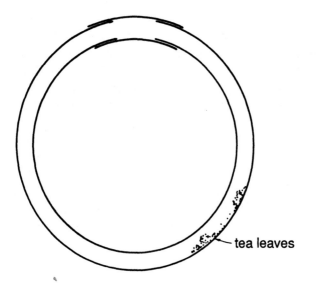

tea leaves

Figure A.2. A ring of plastic tubing containing water and tealeaves.

We marked the torus in octants. There is a motor-driven table at Woods Hole that can be adjusted to go around once every 8.5 seconds (10,000 times the rate of the earth), so one of us sat on it, instead of trying to twirl around uniformly while standing on the floor. When the tealeaves seemed to be at rest in the rotating frame we'd turn the torus over and timed the tealeaves as they passed through one or more octants. The time was crudely determined by saying "mark" twice, and using a stopwatch. The results of these trials are shown in table A.1. The ratios of times for the tealeaves to traverse n octants to the times for the table to traverse n octants in absolute (really terrestrial) space are shown on the bottom line of the table. They decrease as n increases. This is to be expected because of the turbulent dissipation in the torus. Evidently for $n = 0$ the ratio must be close to 2, the factor of our crude formula (A.7), with $\vartheta = \pi/2$. This is a crude demonstration by Compton's standards, but could be refined.

TABLE A.1. Times in seconds for tealeaves to traverse n octants after inversion of the torus.

OCTANTS n	1	2	4
	.6	1.4	3.2
	.8	1.6	3.5
	.5	1.1	2.8
	.6	1.4	2.9
Mean	.63	1.38	3.10
Time for table	1.06	2.12	4.25
Ratio	1.7	1.5	1.4

A.4 THE COMPTON GENERATOR

Because one of the drawbacks of trying to do this experiment quantitatively is the rapid decay of the motion of the water in the tube due to internal turbulent dissipation after it has been inverted, a static experiment would seem to be called for.

Consider the possibility of building an induction generator that would demonstrate the earth's rotation (in a variation of Compton's experiment) by maintaining a small steady difference of level (observable in a manometer) rather than inducing a temporary decaying velocity of flow. In figure A.3A a short cylinder is wrapped with 10 turns of 1/4 inch plastic tubing and fixed to a horizontal axis so that it can be turned with a crank. The free ends of the coil of tubing are connected to the half of a commutator fixed on the shaft, while the other half is held stationary and connected to a manometer. The holes in the commutator are small, so that that it is closed except when the axis of the coils is horizontal, at which point they are briefly open. A colleague, John Thomson, kindly built one out of scrap material in the shop, and it performs handsomely. It is shown in plate 1. A simplified one turn version is shown in figure A.3B.

The apparatus may now be clamped to the seat of a rotatable drafting stool. There are two degrees of rotation available in this system: (1) *turning* about the horizontal axis by means of the hand crank, and (2) *rotation* of the entire apparatus about a vertical axis by pushing the apparatus around. For convenience of differentiating between them we will speak of (1) as *turning* and (2) as *rotating*. If you do either of these separately to the apparatus the levels in the two manometer tubes will remain the same. But if you do both together there will be a no-

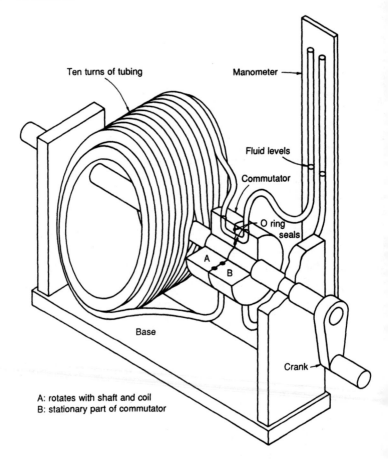

Ten turns of tubing

Manometer

Fluid levels

Commutator

O ring seals

A

B

Base

Crank

A: rotates with shaft and coil
B: stationary part of commutator

Figure A.3.A A cutaway sketch showing the construction of the Compton generator, particularly of the commutator. The valves in the commutator are positioned so that they are open only when the coil of turns in a vertical plane, thus communicating to the manometer only at the moment of maximum of the Coriolis force.

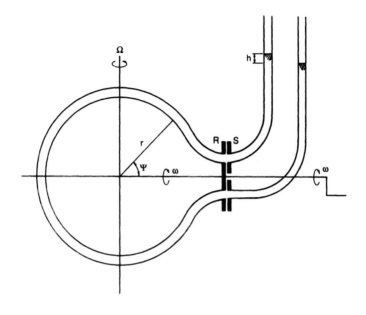

Figure A.3.B A simplified Compton generator of one turn coil. The ends of the coil connect to the manometer at the left through a fluid-valve type commutator. The side of the valve that rotates, *R*, is fixed to the horizontal shaft that is turning with angular velocity ω. The side of the valve that is stationary, *S*, is connected to the manometer. The entire apparatus is mounted on a rotating table so that it can be rotated around a vertical axis with angular velocity Ω.

Because of the arrangement of the valves, they are open only at the moment when the plane of the coil is vertical, and the top end of the coil is always connected to the lefthand tube of the manometer at this moment for transmission of fluid or fluid pressure.

Plate 1. The Compton Generator. Photograph of the Compton generator, showing the crank by which the coil may be turned about a horizontal axis. The entire apparatus may be clamped to the top of a rotatable draughting stool, or piano stool, so that it can be turned about a vertical axis at the same time as the crank is turned about a horizontal axis.

ticeable difference in level. For example, suppose that you walk slowly around the stool so that it takes 8.5 seconds to make one rotation, and turn the crank at 1 c.p.s., you will observe a difference of water level of about 40 cm. in this homely apparatus. There is similarity to a D.C. electrical generator, with the earth's rotation standing for a fixed magnetic field, the Coriolis force standing in for e.m.f., pressure in the fluid standing in for voltage, and the fluid flow as electrical current. For reasons of the analogy we call our machine a Compton generator. The static signal is surprisingly big and suggests that with 100 turns, three times larger diameter coil, and turning of the crank by motor at 10 c.p.s., one might dispense with the stool altogether and measure the earth's rotation. But one is likely to be defeated by the flexibility of the tubes, the compressibility of water, the design of the commutator, and many complications that might make it very inefficient.

The first order theory of this system is simple. It will be instructive to obtain it in two different ways: (1) in the rotating reference frame using Coriolis forces resolved along the tubing, (2) in absolute space by consideration of the direction of the centrifugal force on a particle fixed in the tube (we will need to find the direction of the osculating plane, amplitude of velocity, and curvature of trajectory at the moment when the commutator is open). By comparing these two methods we will discover the computational advantages of doing the problem in the rotating reference frame. The more difficult (computationally) method of the osculating plane in absolute space gives a strong Newtonian flavor to the problem, is physically illuminating, and intuitively resonant with first principles.

Let us imagine that the Compton generator is operating with static head in the manometer, so there is no fluid flow. We also imagine that because the tube and water are effectively incompressible the pressure is at all times balanced by the components of the driving forces along the tubing of the coils. The pressure in the coils reverses periodically as the crank is turned. It is therefore sufficient to consider what these forces are at the time of maximum, because that is what the commutator permits the manometer to measure.

A.5 COMPUTATION (1).
ROTATING REFERENCE FRAME.

This calculation is already done in a more general way in equations (A.1–A.7). Let us take the angular frequency of *turning* about the horizontal axis to be ω and that of *rotation* of the stool about the vertical axis to be Ω, in radians per second, and the radius of the coils as r. Assuming that the pressure in the tubing of the coil is always balanced by the component of the driving force tangential to the tubing and that the commutator valves are open only at the instant when the plane of the coil is vertical and the Coriolis force a maximum, then the maximum x-component of Coriolis force per unit mass at P from expression A.3 is $2\Omega v_{max}$ where v_{max} is the maximum value of v at the top of the coil. The integral of the component tangential to the tubing in the coil of ten turns is then balanced by the head h in the manometer

$$gh = 10(2\Omega v_{max}) \int_0^{2\pi} \sin^2\psi \, d\psi = 20\pi\Omega r v_{max}. \quad (A.8)$$

With $\omega = 2\pi \ s^{-1}$ and $r = 10$ cm then $v_{max} = 72.3$ cm s^{-1}. With $\Omega = 2\pi/8.5$ and $g = 980$ cm s^{-2} we find

$$h = \frac{1}{980}(200 \cdot \pi \cdot 72.3/8.5) = 35 \, cm.$$

A.6 COMPUTATION (2). AS SEEN
IN INERTIAL SPACE

In inertial space, at absolute rest, let us choose the trajectory of the particle to be given by

$$x = f_1(t) \equiv r \cos \vartheta \cos \phi,$$

$$y = f_2(t) \equiv r \cos \vartheta \sin \phi,$$

and

$$z = f_3(t) \equiv r \sin \vartheta, \tag{A.10}$$

where $\dot{r} = 0$, $\phi = \Omega t - \pi/2$, $\vartheta = \omega t + \pi/2$. The particle reaches its highest point $z = r$ at $t = 0$, and you will notice (in fig. A.4) that it lies in the $x < 0$ half-space both before and after this event. Therefore the osculating plane of the trajectory at the pole (the highest point) is tilted, and the centrifugal force must be directed within the osculating plane along a line somewhat different from that of the z-axis, with an x-component. We want to show that this simple geometry of centrifugal force gives the same result as Computations (1) and (2). The geometry is more complicated, but the physics clearer.

Using Taylor's Theorem to obtain values of the functions f_i at three points, in order to determine the osculating plane at $x = y = 0$, $z = r$ we obtain

$$\begin{vmatrix} x & y & z-r \\ \dot{f}_1 & \dot{f}_2 & \dot{f}_3 \\ \ddot{f}_1 & \ddot{f}_2 & \ddot{f}_3 \end{vmatrix} = 0, \tag{A.11}$$

which upon substitution of the explicit expressions for the f_i functions becomes

$$\dot{\vartheta} x = 2 \dot{\phi}(z - r), \tag{A.12}$$

or

$$z = r + \frac{\omega}{2\Omega} x. \tag{A.13}$$

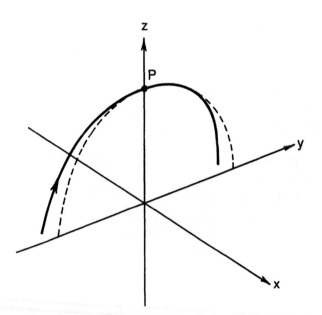

Figure A.4. The solid curve shows the circle that osculates the trajectory (in inertial space) of a particle of fluid at rest with respect to the tube at point *P*. The dashed curve shows the trajectory of the particle were $\Omega = 0$.

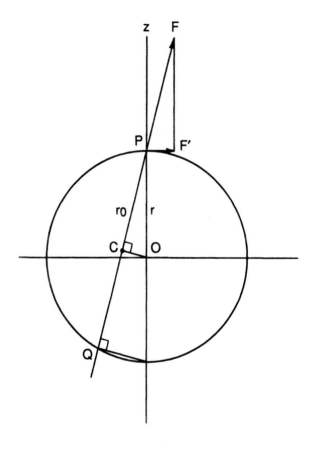

Figure A.5. The centrifugal force *PF* in the osculating plane *QCP* (normal to the plane of the figure) has a component *PF'* tangential to the ring of the tubing (in the plane of the figure).

The center of curvature (with radius r_0) of the osculating circle lies in the x, z plane at point C (figure A.5), so the following ratio obtains

$$\frac{r}{r_0} = \frac{\overline{PF}}{\overline{F'F}} = \frac{\overline{PF}}{\overline{PF'}} \frac{2\Omega}{\omega}, \qquad (A.14)$$

where $\overline{PF}$ is the instantanous value of the centrifugal force (and $\overline{PF'}$ is the x-component of this force that we are seeking) acting on the particle at $t = 0$. We can obtain this by noting that the particle's velocity

$$\sqrt{(r\cos\vartheta\,\dot\phi)^2 + (r\dot\vartheta)^2} = r\dot\vartheta = r\omega$$

at $t = 0$. Thus

$$\overline{PF} = \frac{r^2\omega^2}{r_0}. \qquad (A.15)$$

Now substituting (A.15) into (A.13) we find $\overline{PF'} = 2\Omega\omega r$ which is the same expression as before. So now we have given a more physically intuitive explanation of the Compton generator. We might even say that it is a form of centrifugal pump.

A.7 COMPTON SAVES HIMSELF

The reader will still be nagged by the question of how Compton resolved the enigma of his first paper. So far as we can determine from the second paper reporting the repetition of the experiment at Princeton, we do not see that he did formally address the theoretical question or refer to the first paper. Instead, as a superlative experimenter, Compton sought the high ground of experimental ingenuity, and managed to bypass the theoretical question. In one series of experiments he mounted his apparatus on a slowly rotatable spectrograph table (presumably belonging to the astronomy department). He adjusted the clockwork of the table until the fluid in his torus rotated, after inverting it, with exactly equal magnitude but opposite sign to that observed with the clockwork stopped. Clearly under these circumstances the spectrograph table was rotating in the opposite direction from that of the earth in inertial space. The magnitude of the vertical component of the earth's rotation is thus half that of the spectrograph table's rate relative to the earth. The other observed quantities—latitude and direction of north relative to the laboratory apparatus—depend only upon ratios. Thus Compton managed to get his results without appeal to theory in his Princeton experiment. Within a few years he won the Nobel Prize in physics, became the chairman of the physics department at the University of Chicago and the Chancellor of Washington University. One wonders, just the same, whether this truly great experimental physicist ever completely conquered his boyhood vertigo when thinking about rotating systems.

EXERCISES

Exercise A-1 [lines 55000–55250] *Osculating plane.* This program illustrates the motion in absolute space of the point *P* that is at the top of the coil at time $t = 0$, from a time before, when it is as the "equator" to any future time. The OMEGA [55030] is the angular rotational velocity of the stool, the RATIO is the ratio of the angular turning velocity of the crank around the x-axis to the angular rotational velocity OMEGA. Initial values of absolute latitude LA(1) and longitude LO(1) of the point are fixed [55040]. Computation during other times is made in lines 55060−70, as well as of the absolute latitude LA(2) and longitude LO(2) of a point on a meridian at time $t = 0$. Both are converted to plotting coordinates in 55080−55100, and plotted by 55110, and then the loop is repeated [55120]. The subroutine [55130] draws the grid of the sphere that circumscribes the ring of the coil in absolute space.

```
55000 'Exercise A-1; osculating plane      **
55010 SCREEN 1: COLOR 0,2: KEY OFF: CLS
55020 PI = 3.14159: OMEGA=10:DT =.0003
55021 FACT = 60:INC=PI/8:K=PI/4
55030 RATIO = 8.5: N =RATIO*OMEGA
55040 LA(1) = 0 : LO(1)=-PI/2-K
55050 GOSUB 55130
55060 LO(1)=OMEGA *DT+LO(1)
55070 LA(1)= N*DT+LA(1) :T=T+DT :LA(2)=LA(1)
55071 LO(2)=-PI/2-K +PI/(2*RATIO)
55080 FOR I = 1 TO 2: RHO(I)=COS(LA(I))
55090 X(I)=RHO(I)*COS(LO(I))
55091 Y(I)=RHO(I)*SIN(LO(I))*SIN(INC)
55100 Z(I)=Y(I)+SIN(LA(I))*COS(INC)
55110 PSET(70+FACT*X(I),90-FACT*Z(I)),I+1
55111 NEXT
55120 GOTO 55060
55130 J = INC: ' plot the two spheres "
55140 I = 1 :SS=SIN(J)
55150 CIRCLE (220-150*I,90),60,1,0,2*PI,1
55160 CIRCLE (220-150*I,90),60,1,0,2*PI,SS
55170 C9=60*SIN(PI/6)*COS(J)
55180 K9=60*COS(PI/6)
55190 CIRCLE (220-150*I,90-C9),K9,1,0,2*PI,SS
55200 CIRCLE (220-150*I,90+C9),K9,1,0,2*PI,SS
55210 C9=60*SIN(PI/3)*COS(J)
55220 K9=60*COS(PI/3)
55230 CIRCLE(220-150*I,90-C9),K9,1,0,2*PI,SS
55240 CIRCLE(220-150*I,90+C9),K9,1,0,2*PI,SS
55250 RETURN
```

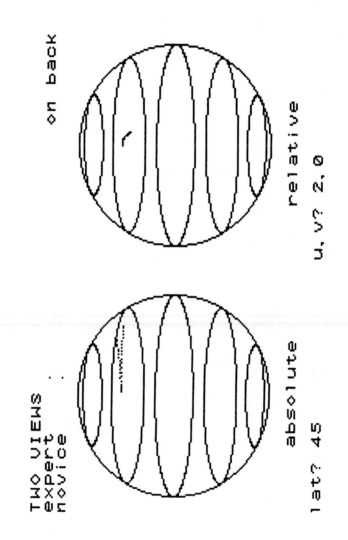

TWO VIEWS
expert :
novice

absolute   on back

relative

lat? 45   u, v? 2, 0

SAMPLE VIEW OF THE SCREEN:
Example 7-1

This is what the screen should look like when using the program called Exercise 7-1. The display on the left is in absolute inertial space, on the right with reference to the rotating earth. After entering [as a sample case] Lat = 45 and u, v = 2, 0, the particle's trajectory starts being drawn in each reference frame.

With some computer configurations there is a distortion in the graphics, tending to stretch the picture in the vertical, amounting to a factor of about 6/5. This can often be simply remedied by insertion of a WINDOW SCREEN instruction early in the program. If the vertical distortion is present but doesn't disturb you, then remember that, if uncorrected for, it will affect all the displays. For example, plane views of rotating dishes and inertial circles will be similarly distorted.

Index

CPSIA information can be obtained at www.ICGtesting.com
Printed in the USA
BVOW07s1637131113

336203BV00001B/1/P

9 780231 066372